BOUNDARY VALUE PROBLEMS

AND PARTIAL

DIFFERENTIAL EQUATIONS

Student Solutions Manual

FIFTH EDITION

DAVID L. POWERS

ELSEVIER
ACADEMIC
PRESS

AMSTERDAM • BOSTON • HEIDELBERG • LONDON
NEW YORK • OXFORD • PARIS • SAN DIEGO
SAN FRANCISCO • SINGAPORE • SYDNEY • TOKYO

Acquisitions Editor: Tom Singer
Project Manager: Jeff Freeland
Sr. Marketing Manager: Linda Beattie
Cover Design: Eric DeCicco
Cover Printer: LexisNexis
Interior Printer: LexisNexis

Elsevier Academic Press
30 Corporate Drive, Suite 400, Burlington, MA 01803, USA
525 B Street, Suite 1900, San Diego, California 92101-4495, USA
84 Theobald's Road, London WC1X 8RR, UK

ISBN-13: 978-0-12-088586-2
ISBN-10: 0-12-088586-7

For all information on all Elsevier Academic Press publications
visit our Web site at www.books.elsevier.com

Printed and bound in the United Kingdom

Transferred to Digital Print 2011

TABLE OF CONTENTS

Chapter 0

0.1 Homogeneous Linear Equations

1. You should be able to write out the solution without going through any algebra

$$\phi(x) = c_1 \cos(\lambda x) + c_2 \sin(\lambda x).$$

3.(a) Treat this as a constant-coefficients equation. The characteristic equation is $m^2 = 0$, with double root $m = 0$. Therefore the solution of the differential equation is $u(t) = c_1 + c_2 t$.

(b) Because there is no u or du/dt term, you can integrate directly, twice: $du/dt = c_2$, $u = c_2 t + c_1$.

5. Do the indicated differentiation.

$$\frac{d^2 W}{dr^2} + \frac{1}{r}\frac{dW}{dr} - \frac{\lambda^2}{r^2} W = 0.$$

This is a Cauchy-Euler equation. Guess $w = r^m$ so the characteristic equation is $m(m-1) + m - \lambda^2 = 0$ or $m^2 - \lambda^2 = 0$, with solutions $m = \lambda$, $m = -\lambda$. The general solution is

$$w(r) = c_1 r^\lambda + c_2 r^{-\lambda}.$$

7. This differential equation is best solved by integrating (since there is no term in v).

$$(h + kx)\frac{dv}{dx} = c_1$$

$$\frac{dv}{dx} = \frac{c_1}{h + kx}$$

$$v = \frac{c_1}{k}\ln(h + kx) + c_2$$

9. Solve by integrating, since there is no term in u:

$$x^3 \frac{du}{dx} = c_1; \quad \frac{du}{dx} = c_1 x^{-3};$$

$$u = -\frac{c_1}{2} x^{-2} + c_2.$$

11. Solve by integrating, since there is no term in u.

$$r\frac{du}{dr} = c_1; \quad \frac{du}{dr} = \frac{c_1}{r}; \quad u = c_1 \ln(r) + c_2.$$

13. The characteristic polynomial is $(m^4 - \lambda^4) = (m^2 + \lambda^2)(m^2 - \lambda^2)$ with roots $m = \pm\lambda$, $\pm i\lambda$. The general solution of the differential equation is

$$u(x) = c_1 e^{\lambda x} + c_2 e^{-\lambda x} + c_3 \cos(\lambda x) + c_4 \sin(\lambda x).$$

15. Guess $u = e^{mx}$. Then the characteristic equation is $m^4 + 2\lambda^2 m^2 + \lambda^4 = 0$. This is a biquadratic polynomial equation: $(m^2 + \lambda^2)^2 = 0$. The roots are $m = \pm i\lambda$, both having multiplicity 2. The general solution is

$$u = (c_1 + c_2 x)\cos(\lambda x) + (c_3 + c_4 x)\sin(\lambda x).$$

17. Assume $u_2 = t^b \cdot v$. Then

$$u_2' = bt^{b-1}v + t^b v'; \quad u_2'' = b(b-1)t^{b-2}v + 2bt^{b-1}v' + t^b v''.$$

Substituting gives

$$b(b-1)t^b v + 2bt^{b+1}v' + t^{b+2}v'' + (1-2b)(bt^b v + t^{b+1}v') + b^2 t^b v = 0.$$

Sort terms by derivatives

$$t^{b+2}v'' + (2bt^{b+1} + (1-2b)t^{b+1})v' + \left(b(b-1) + b(1-2b) + b^2\right)t^b v = 0.$$

The coefficient of v has to be 0. If not, there was an error. The differential equation reduces to $tv'' + v' = 0$, which is a Cauchy-Euler equation. The characteristic equation is $m^2 = 0$, with double root $m = 0$. Thus $v(t) = c_1 + c_2\ln(t)$ and the general solution of the original equation is $u(t) = (c_1 + c_2\ln(t))\, t^b$. Choosing $c_1 = 0$, $c_2 = 1$ gives a second solution.

19. Replace each R by u/ρ.

$$\frac{d}{d\rho}\left(\rho^2 \frac{d}{d\rho}\left(\frac{u}{\rho}\right)\right) + \lambda^2\rho^2\frac{u}{\rho} = 0.$$

Differentiate $\dfrac{d}{d\rho}\left(\dfrac{u}{\rho}\right) = \dfrac{\rho u' - u}{\rho^2}$ substitute and clean up

$$\frac{d}{d\rho}\left(\rho\frac{du}{d\rho} - u\right) + \lambda^2\rho u = 0.$$

Differentiate

$$\rho\frac{d^2 u}{d\rho^2} + \frac{du}{d\rho} - \frac{du}{d\rho} + \lambda^2\rho u = 0.$$

Simplify

$$\frac{d^2 u}{d\rho^2} + \lambda^2 u = 0.$$

The equation comes up in physical problems of waves in 3d.

21. Use the chain rule:

$$\frac{du}{dt} = \frac{dv}{dx} \cdot \frac{dx}{dt} = \frac{dv}{dx} \cdot \frac{1}{t}$$

$$\frac{d^2 u}{dt^2} = \frac{d}{dt}\left(\frac{dv}{dx}\cdot\frac{1}{t}\right) = \frac{dv}{dx}\left(-\frac{1}{t^2}\right) + \frac{d}{dx}\left(\frac{dv}{dx}\right)\cdot\frac{1}{t^2}.$$

Now substitute:

$$\frac{d^2 v}{dx^2} - \frac{dv}{dx} + k\frac{dv}{dx} + pv = 0.$$

This is a constant-coefficient equation with characteristic equation $m^2 + (k-1)m + p = 0$.

23. The characteristic equation is $m^2 + 2\alpha m + \sigma^2 = 0$, with roots $m = -\alpha \pm \sqrt{\alpha^2 - \sigma^2}$. We know that the system is underdamped, so $\sigma^2 > \alpha^2$ and we set $\sqrt{\alpha^2 - \sigma^2} = i\beta$. The solution of the differential equaiton is

$$y(t) = e^{-\alpha t} \left(c_1 \cos(\beta t) + c_2 \sin(\beta t) \right).$$

The initial conditions determine c_1, c_2.

$$y(0) = -.001h : c_1 = -.001h$$

$$y'(0) = 0 : -\alpha c_1 + \beta c_2 = 0$$

The complete solution is

$$y(t) = -.001h \left(\cos(\beta t) + \frac{\alpha}{\beta} \sin(\beta t) \right).$$

25. Damping is critical when

$$\left(\frac{2\alpha}{2} \right)^2 = \sigma^2 \quad \text{or} \quad \alpha/\sigma = 1.$$

This occurs at $v = 2.62 m/s$.

Chapter 0

0.2 Nonhomogeneous Linear Equations

1. In standard form the equation is $u' + au = aT$. The inhomogeneity is aT, assumed constant. Guess $u_p(t) = A$ and substitute to find $0 + aA = aT$. Hence $u_p = T$, and the general solution is $u(t) = T + c_1 e^{-at}$.

3. Guess $u_p(t) = Ae^{-at}$. Substituting in the differential equation gives $-Aae^{-at} + aAe^{-at} = e^{-at}$. This is impossible, since the left-hand side is 0, because Ae^{-at} is a solution of the homogeneous equation. Now apply the revision rule: guess $u_p(t) = Ate^{-at}$. Substituting in the differential equation gives $A(-ate^{-at} + e^{-at}) + aAte^{-at} = e^{-at}$. The two terms containing t cancel, leaving $Ae^{-at} = e^{-at}$. Hence $A = 1$ and the general solution is $u(t) = te^{-at} + ce^{-at}$, with c arbitrary.

5. The solution to guess is $u_p(t) = A\cos(t) + B\sin(t)$. However, this is (for any A, B) a solution of the homogeneous equation $u'' + u = 0$. Hence, we must apply the revision rule and guess $u_p(t) = t(A\cos(t) + B\sin(t))$. Substituting into the differential equation gives $t(-A\cos(t) - B\sin(t)) + 2(-A\sin(t) + B\cos(t)) + t(A\cos(t) + B\sin(t)) = \cos(t)$. Terms containing t cancel; matching coefficients gives $-2A = 0$, $-2B = 1$. Thus the general solution is $u(t) = -\frac{1}{2}t\sin(t) + c_1\cos(t) + c_2\sin(t)$.

7. This is a deceptive equation, because (1) $\cosh(t) = \frac{1}{2}e^t + \frac{1}{2}e^{-t}$ and (2) the general solution of the homogeneous equation is $c_1 e^{-t} + c_2 e^{-2t}$. Thus, we need to break up the cosh, solve for two particular solutions, and use the revision rule for one of them.

To find the particular solution of $u'' + 3u' + 2u = \frac{1}{2}e^t$, guess $u_{p1}(t) = Ae^t$ and find $(A + 3A + 2A)e^t = \frac{1}{2}e^t$, so $A = 1/12$. To find the particular solution of $u'' + 3u' + 2u = \frac{1}{2}e^{-t}$, apply the revision rule and guess $u_{p2}(t) = Bte^{-t}$. We find

$$B(te^{-t} - 2e^{-t}) + 3B(-te^{-t} + e^{-t}) + 2Bte^{-t} = \frac{1}{2}e^{-t}.$$

The terms with t drop out, leaving $(-2 + 3)Be^{-t} = \frac{1}{2}e^{-t}$, so $B = \frac{1}{2}$. Finally, the general solution is

$$u(t) = \frac{1}{12}e^t + \frac{1}{2}te^{-t} + c_1 e^{-t} + c_2 e^{-2t}.$$

9. Recall that the method of undetermined coefficients is not guaranteed to work if the differential equation has variable coefficients, as this one has. This equation is best solved by algebra and integration:

$$(\rho^2 u')' = -\rho^2; \quad \rho^2 u' = -\frac{1}{3}\rho^3 + c_1;$$

$$u' = -\frac{1}{3}\rho + c_1\rho^{-2}; \quad u = -\frac{1}{6}\rho^2 - c_1\rho^{-1} + c_2.$$

11. Divide through by M, and let $K/M = b$. Then

$$\frac{d^2 h}{dt^2} + b\frac{dh}{dt} = -g.$$

The equation is linear, **non**homogeneous, with constant coefficients. The characteristic equation is $m^2 + bm = 0$ with roots $-b$ and 0. Thus $h_c(t) = c_1 e^{-bt} + c_2$. Because the r.h.s. is constant, **and** constant is a solution of the homogeneous equation, the guessed form for the particular solution is (see p. 16) $h_p(t) = At$ (not just A).

The general solution of the differential equation is

$$h(t) = c_1 e^{-bt} + c_2 - \frac{gt}{b}.$$

The initial conditions require

$$c_1 = -\frac{g}{b^2}, \quad c_2 = h_0 + \frac{g}{b^2},$$

and finally

$$h(t) = h_0 - \frac{gt}{b} + \frac{g}{b^2}\left(1 - e^{-bt}\right).$$

13. Seek a solution in the form $u(t) = v(t)e^{-at}$. Substitute to find

$$v(-ae^{-at}) + v'e^{-at} + ave^{-at} = e^{-at},$$

which leads to $v' = 1$ or $v(t) = t$. Thus $u_p(t) = te^{-at}$.

15. From Eq. (12') and (15) we have

$$\left.\begin{array}{l} v_1' \cos(x) + v_2' \sin(x) = 0 \\[2em] -v_1' \sin(x) + v_2' \cos(x) = \tan(x) \end{array}\right\}$$

with solution

$$v_1'(x) = -\sin^2(x)/\cos(x), \quad v_2'(x) = \sin(x).$$

Change $v_1'(x)$ to $(\cos^2(x) - 1)/\cos(x) = \cos(x) + \sec(x)$ to integrate. Then $v_1(x) = \sin(x) - \ln|\tan(x) + \sec(x)|$, $v_2(x) = -\cos(x)$. This problem could *not* be solved by undetermined coefficients.

17. The system of equations to solve is

$$\left.\begin{array}{r} v_1' + v_2't = 0 \\[1em] v_2' = -1 \end{array}\right\}$$

with solution $v_2' = -1, v_1' = t$. Hence $v_1 = t^2/2$, $v_2 = -t$ and $u_p(t) = t^2/2 - t^2 = -t^2/2$. The same solution could be found by integration.

19. Note that the differential equation has to be divided by t^2 to be in standard form. The system of equations to solve is

$$\left.\begin{array}{r} v_1't + v_2'/t = 0 \\[1em] v_1' - v_2'/t^2 = 1/t^2 \end{array}\right\}$$

with solution $v_1' = 1/2t^2$, $v_2' = -1/2$. Hence $v_1 = -1/2t$, $v_2 = -t/2$ and $u_p(t) = -1/2 - 1/2 = -1$.

21. Here we must follow the development of the theorem. Assume $u_p(t) = e^{-at}v(t)$. Then $(e^{-at}v' - ae^{-at}v) + ae^{-at}v = f(t)$ or $e^{-at}v' = f(t)$. Because the equation is first order, there is only one equation for v'. The solution is $v' = e^{at}f(t)$, from which $v(t) = \int_0^t e^{az}f(z)dz$ and $u_p(t) = e^{-at}\int_0^t e^{az}f(z)dz$.

23. To get the multiplier α out of the parentheses, choose $\beta = 1/\alpha$. Then collect multipliers to find $K = R\alpha/\rho c$.

25. The easiest way to solve is by integration. Divide through the equation by $(\beta + T)$:

$$\frac{1}{\beta + T} \frac{dT}{dt} = KI_{max}^2 e^{-2\lambda t}$$

$$\ln(\beta + T) = KI_{max}^2 \frac{e^{-2\lambda t}}{-2\lambda} + c.$$

From the initial condition, find

$$c = \ln\beta + KI_{max}^2/2\lambda.$$

Then $\ln(\beta + T) = \ln\beta + KI_{max}^2(1 - e^{-2\lambda t})/2\lambda$. Now, use each side as an exponent:

$$\beta + T = \beta \exp\left(KI_{max}^2(1 - e^{-2\lambda t})/2\lambda\right).$$

Subtract β from each side to get $T(t)$.

Chapter 0

0.3 Boundary Value Problems

1.a. The solution of the differential equation is $u(x) = c_1 \cos(x) + c_2 \sin(x)$. Applying the boundary conditions gives two equations:

$$u(0) = 0: \quad c_1 = 0$$

$$u(\pi) = 0: \quad -c_1 = 0$$

Therefore $c_1 = 0$ and c_2 is arbitrary, giving infinitely many solutions.

b. Solution $u(x) = 1 + c_1 \cos(x) + c_2 \sin(x)$. Boundary conditions:

$$u(0) = 0: \quad 1 + c_1 = 0$$

$$u(1) = 0: \quad 1 + c_1 \cos(1) + c_2 \sin(1) = 0$$

The solution is $c_1 = -1$, $c_2 = \dfrac{-1 + \cos(1)}{\sin(1)}$. Unique solution.

c. Solution $u(x) = c_1 \cos(x) + c_2 \sin(x)$. Boundary conditions:

$$u(0) = 0: \quad c_1 = 0$$

$$u(\pi) = 1: \quad -c_1 = 1$$

These are inconsistent, so no solution is possible.

3. In all cases, the general solution of the differential equation is $u(x) = c_1 \cos(\lambda x) + c_2 \sin(\lambda x)$ if $\lambda \neq 0$. The two boundary conditions give two simultaneous equations for c_1 and c_2.

a. $c_1 = 0$; $-\lambda c_1 \sin(\lambda a) + \lambda c_2 \cos(\lambda a) = 0$. Therefore $\cos(\lambda a) = 0$ and $\lambda = \pm \pi/2a, \pm 3\pi/2a, \pm 5\pi/2a, \cdots$.

b. $\lambda c_2 = 0$; $c_1 \cos(\lambda a) + c_2 \sin(\lambda a) = 0$. Therefore $\cos(\lambda a) = 0$; solutions as in a.

c. $\lambda c_2 = 0$; $-\lambda c_1 \sin(\lambda a) + \lambda c_2 \cos(\lambda a) = 0$. Therefore $\sin(\lambda a) = 0$ and $\lambda = \pm \pi/a, \pm 2\pi/a, \pm 3\pi/a, \cdots$. For this case $\lambda = 0$ is a possibility as well, with solution $u(x) = c_1$.

5. The two boundary conditions become:

$$u(0) = h: \quad c' + \frac{1}{\mu} \cosh(\mu c) = h.$$

$$u(a) = h: \quad c' + \frac{1}{\mu} \cosh(\mu(c + a)) = h.$$

Deduce that $\cosh(\mu c) = \cosh(\mu(c + a))$. By the symmetry of the cosh function, $c + a = -c$, or $c = -a/2$. Then $c' = h - \frac{1}{\mu} \cosh(\mu a/2)$.

7. Let $hC/kA = \gamma^2$. The general solution of the differential equation is $u(x) = T + c_1 \cosh(\gamma x) + c_2 \sinh(\gamma x)$, and $u'(x) = c_1 \gamma \sinh(\gamma x) + c_2 \gamma \cosh(\gamma x)$. Apply the boundary conditions:

$$u(0) = T_0: \quad T_1 + c_1 = T_0$$

$$\kappa u'(a) + hu(a) = hT:$$

$$c_1(\kappa\gamma \sinh(\gamma a) + h \cosh(\gamma a)) + c_2(\kappa\gamma \cosh(\gamma a) + h \sinh(\gamma a)) = 0$$

(hT is on both sides, drops out). See Answers.

9. Let $hC/\kappa A = \gamma^2$, $I^2 R/\kappa A = H$, so

$$u'' - \gamma^2 u = -\gamma^2 T - H, \quad 0 < x < a$$

Guess the particular solution $u_p = B$ (const). Then $B = T + H/\gamma^2$. Also $u_c(x) = c_1 \cosh(\gamma x) + c_2 \sinh(\gamma x)$. The general solution of the differential equation is

$$u(x) = T + \frac{H}{\gamma^2} + c_1 \cosh(\gamma x) + c_2 \sinh(\gamma x).$$

Apply the boundary conditions

$$u(0) = T: \quad T + \frac{H}{\gamma^2} + c_1 = T,$$

or

$$c_1 = -\frac{H}{\gamma^2}$$

$$u(a) = T: \quad T + \frac{H}{\gamma^2} - \frac{H}{\gamma^2} \cosh(\gamma a) + c_2 \sinh(\gamma a) = T,$$

or

$$c_2 = \frac{H}{\gamma^2} \frac{\cosh(\gamma a) - 1}{\sinh(\gamma a)}.$$

Final solution

$$u(x) = T + \frac{H}{\gamma^2}\left[1 - \cosh(\gamma x) + \frac{\cosh(\gamma a) - 1}{\sinh(\gamma a)} \sinh(\gamma x)\right].$$

Sketch this function. Does it belly up or down?

11.

$$\frac{d^2 u}{dy^2} = \frac{-g}{\mu}, \quad 0 < y < L$$

$$u(0) = 0, \quad u(L) = 0.$$

Integrate twice to solve the differential equation

$$u' = \frac{-g}{\mu} y + c_1, \quad u = \frac{-g}{\mu} \frac{y^2}{2} + c_1 y + c_2.$$

Apply boundary conditions

$$u(0) = 0: \quad c_2 = 0$$

$$u(L) = 0: \quad \frac{-g}{\mu}\frac{L^2}{2} + c_1 L + c_2 = 0$$

$$c_1 = \frac{g}{\mu}\frac{L}{2}$$

and

$$u(y) = \frac{g}{\mu}\frac{y(L-y)}{2}.$$

Sketch.

13. The solution of the problem in Exercise 12. is $u(x) = A + Bx + Cx^2 + c_1 \cos(\lambda x) + c_2 \sin(\lambda x)$, where

$$C = \frac{w}{2P}, \quad B = -\frac{wL}{2p}, \quad A = -\frac{EIw}{p^2}, \quad \lambda = \sqrt{\frac{P}{EI}}.$$

Applying the boundary conditions gives $c_1 = -A$, and $C_2 \sin(\lambda L) = 0$. If $\sin(\lambda L) \neq 0$, then $c_2 = 0$, and the solution is unique. If $\sin(\lambda L) = 0$, then c_2 can have any value, and there are infinitely many solutions. The value of λ determines P.

15. Let $hC/\kappa A = \gamma^2$. The differential equation is

$$u'' - \gamma^2 u = -\gamma^2 T - g/\kappa.$$

The general solution of the differential equation is

$$u(x) = T + \frac{g}{\gamma^2 \kappa} + c_1 \cosh(\gamma x) + c_2 \sinh(\gamma x)$$

Apply the b.c. to get the solution in the Answers. Does the graph of $u(x)$ bulge up or down?

17. The general solution of the differential equation is $u_1(r) = c_1 \ln(r) + c_2$, and $u'(r) = c_1/r$. The two boundary conditions are

$$\left.\begin{array}{l} -\dfrac{\kappa c_1}{a} = h_0(T_w - c_1 \ln(a) - c_2) \\[2mm] \dfrac{\kappa c_1}{b} = h_1(T_a - c_1 \ln(b) - c_2) \end{array}\right\}$$

or

$$\left.\begin{array}{l} c_1\left(h_0 \ln(a) - \frac{\kappa}{a}\right) + c_2 h_0 = h_0 T_w \\[2mm] c_1\left(h_1 \ln(a) + \frac{\kappa}{b}\right) + c_2 h_1 = h_1 T_a \end{array}\right\}$$

The solution is

$$c_1 = \frac{h_0 h_1(T_w - T_a)}{h_1 h_0 \ln(b/a) - \kappa(\frac{1}{a} + \frac{1}{b})}$$

$$c_2 = \frac{h_1 h_0(T_a \ln(a) - T_w \ln(b)) - \kappa(\frac{h_0}{b} + \frac{h_1}{a})}{h_1 h_0 \ln(\frac{b}{a}) - \kappa(\frac{1}{a} + \frac{1}{b})}$$

19. Because w is constant, the general solution of the differential equation is

$$u(x) = \frac{w}{EI} \frac{x^4}{24} + c_1 x^3 + c_2 x^2 + c_3 x + c_4.$$

Apply the conditions at $x = 0$;

$$u(0) = 0 : \quad c_4 = 0$$

$$u'(0) = 0 : \quad c_3 = 0$$

Now $u(x) = \frac{w}{EI} \frac{x^4}{24} + c_1 x^3 + c_2 x^2$. The two remaining conditions are

$$u''(a) = 0 : \quad \frac{w}{EI} \frac{a^2}{2} + 6c_1 a + 2c_2 = 0$$

$$u'''(a) = 0 : \quad \frac{w}{EI} a + 6c_1 = 0$$

Giving

$$c_1 = \frac{-w}{EI} \frac{a}{6}, \quad c_2 = \frac{w}{EI} \frac{a^2}{4} \quad \text{and} \quad u(x) = \frac{w}{EI} \left(\frac{x^4}{24} - \frac{ax^3}{6} + \frac{a^2 x^2}{4} \right).$$

Chapter 0

0.4 Singular Boundary Value Problems

1. See the Answers.

3. Multiply by ρ^2 and integrate once to get

$$\rho^2 \frac{du}{d\rho} = -\frac{H}{\kappa} \frac{\rho^3}{3} + c_1.$$

Divide through by ρ^2 and integrate:

$$u(\rho) = -\frac{H}{\kappa} \frac{\rho^2}{6} - \frac{c_1}{\rho} + c_2.$$

Since $u(\rho)$ must be bounded as $\rho \to 0+$, we must have $c_1 = 0$. The second b.c. requires

$$-\kappa \left[-\frac{H}{\kappa} \frac{c}{3} \right] = h \left(-\frac{H}{\kappa} \frac{c^2}{6} + c_2 - T \right).$$

Solve for c_2. Remember that c and all other letters are given constants.

5. The equation for $v = \rho u$ is $v'' + \mu^2 v = 0$, with familiar general solution. Therefore we find

$$u(\rho) = \frac{1}{\rho}(c_1 \cos(\mu\rho) + c_2 \sin(\mu\rho)).$$

Since $u(\rho)$ must be bounded as $\rho \to 0$, $c_1 = 0$. The condition at $\rho = a$ is $u(a) = 0$, or $c_2 \sin(\mu a)/a = 0$. Choosing $c_2 = 0$ gives $u(\rho) \equiv 0$. However if $\mu a = \pi, 2\pi, 3\pi, \cdots$ then $\sin(\mu a) = 0$ and c_2 can be any number. For these values, there is a nonzero solution.

7. The solution of the problem in # 6 is $u(r) = T_0 + g(a^2 - r^2)/4\kappa$. Then $u(0) = T_0 + ga^2/4\kappa$. See Answers.

9. The general solution of the differential equation is $u(x) = -AL^2 e^{-x/L} + c_1 x + c_2$. If u is to be bounded as $x \to \infty$, $c_1 = 0$. Then the condition at $x = 0$ gives $c_1 = AL^2 + T_0$, so $u(x) = AL^2(1 - e^{-x/L}) + T_0$.

Chapter 0

0.5 Green's Functions

1. $u_1(x) = x, \quad u_2(x) = x - a,$

$$W(x) = \begin{vmatrix} x & x - a \\ 1 & 1 \end{vmatrix} = a$$

$$G(x, z) = \begin{cases} z(x-a)/a & 0 < z \leq x \\ x(z-a)/a & x \leq z < a \end{cases}$$

3. $u_1(x) = \cosh(\gamma x), \ u_2(x) = \sinh(\gamma(a - x))$

$$W(x) = \begin{vmatrix} \cosh(\gamma x) & \sinh(\gamma(a - x)) \\ \gamma \sinh(\gamma x) & -\gamma \cosh(\gamma(a - x)) \end{vmatrix}$$

$$= -\gamma \left[\cosh(\gamma x) \cosh(\gamma(a - x)) + \sinh(\gamma x) \sinh \gamma(a - x) \right] = -\gamma \cosh(\gamma a)$$

(using addition theorem for cosh. See Mathematical References in the back of the book.)

5.

$$u_1(\rho) = 1, \quad u_2(\rho) = \frac{c}{\rho} - 1 = \frac{c - \rho}{\rho}$$

$$W(\rho) = \begin{vmatrix} 1 & \frac{c-\rho}{\rho} \\ 0 & -\frac{c}{\rho^2} \end{vmatrix} = -\frac{c}{\rho^2}$$

(See Answers).

7. $u_1(x) = \sinh(\gamma x), \quad u_2(x) = e^{-\gamma x}$

$$W(x) = \begin{vmatrix} \sinh(\gamma x) & e^{-\gamma x} \\ \gamma \cosh(\gamma x) & -\gamma e^{-\gamma x} \end{vmatrix} = \gamma e^{-\gamma x}(\sinh(\gamma x) + \cosh(\gamma x)) = -\gamma,$$

using the definition of sinh and cosh. See Answers.

9. Use Equation (18) and the answer to #5

$$
\begin{aligned}
u(x) \;=\;& \int_0^c G(\rho, z) \cdot dz \\
=\;& \int_0^\rho \frac{(c-\rho)/\rho}{-c/z^2}\,dz + \int_\rho^c \frac{(c-z)/z}{-c/z^2}\,dz \\
=\;& \frac{c-\rho}{-c\rho} \int_0^\rho z^2\,dz + \frac{1}{-c} \int_\rho^c (cz - z^2)\,dz \\
=\;& \frac{c-\rho}{-c\rho} \left(\frac{z^3}{3}\Big|_0^\rho \right) - \frac{1}{c}\left(c\frac{z^2}{2} - \frac{z^3}{3}\Big|_\rho^c \right) \\
=\;& \frac{c-\rho}{-c} \left(\frac{\rho^2}{3} \right) - \frac{1}{c}\left(\frac{c^3}{6} - \frac{c\rho^2}{2} + \frac{\rho^3}{3} \right) \\
=\;& -\frac{\rho^2}{3} + \frac{\rho^3}{3c} - \frac{c^2}{6} + \frac{\rho^2}{2} - \frac{\rho^3}{3c} \\
=\;& \frac{\rho^2 - c^2}{6}
\end{aligned}
$$

Direct integration is much easier.

11. The solution $u(x)$ of the differential equation has different formulas: (i) for values of x between 0 and $a/2$, and (ii) for x between $a/2$ and a. In both cases, the first of the two integrals for $u(x)$ (see the Answers) has value 0.

13. See the Answers.

Chapter 0

Miscellaneous

1. See the Answers.

3. $u(x) = c_1 + c_2 x$. Apply the b.c.

$$u(0) = T_0: \quad c_1 = T_0$$

$$u'(a) = 0 : c_2 = 0$$

5. Assume that p is constant. The differential equation becomes

$$(ru')' = -pr; \quad ru' = -\frac{pr^2}{2} + c_1;$$

$$u' = -\frac{pr}{2} + \frac{c_1}{r}; \quad u = -\frac{pr^2}{4} + c_1 \ln(r) + c_2.$$

Boundedness at $r = 0$ requires $c_1 = 0$. Then $u(a) = 0$ gives $-pa^2/4 + c_2 = 0$.

7. Assume that H is constant.

$$(\rho^2 u')' = -H\rho^2; \quad \rho^2 u' = -\frac{H\rho^3}{3} + c_1;$$

$$u' = -\frac{H\rho}{3} + \frac{c_1}{\rho^2}; \quad u = -\frac{H\rho^2}{6} - \frac{c_1}{\rho} + c_2.$$

Boundedness at $\rho = 0$ requires $c_1 = 0$. Then $u(a) = T_0$ gives $-Ha^2/6 + c_2 = T_0$.

9. $u_p(x) = T$ (assumed constant), and

$$u(x) = T + c_1 \cosh(\gamma x) + c_2 \sinh(\gamma(x)).$$

Apply the b.c.: $u'(0) = 0$: $c_2\gamma = 0$; $u(a) = T_1$: $T + c_1 \cosh(\gamma a) = T_1$.

11. $u_p(x) = T_0$ (assumed constant), and $u(x) = T_0 + c_1 e^{\gamma x} + c_2 e^{-\gamma x}$ (use exponentials because of the unbounded region). The requirement that $u(x)$ be bounded as $x \to \infty$ means that $c_1 = 0$. Then $u(0) = T$ gives $T_0 + c_2 = T$.

13. Assume that e is a constant. The differential equation becomes

$$(hh')' = -e; \quad hh' = -ex + c_1; \quad \frac{h^2}{2} = -\frac{ex^2}{2} + c_1 x + c_2.$$

Thus $h(x) = \sqrt{2c_2 + 2c_1 x - ex^2}$. Now apply the b.c.:

$$h(0) = h_0: \quad \sqrt{2c_2} = h_0; \quad c_2 = \frac{h_0^2}{2}.$$

$$h(a) = h_1: \quad \sqrt{h_0^2 + 2c_1 a - ea^2} = h_1;$$

$$c_1 = \frac{(h_1^2 - h_0^2 + ea^2)}{2a}.$$

15. $u_p(x) = \frac{wEI}{k} = H$, and let $\gamma^4 = k/4EI$ for convenience. The characteristic equation for the homogeneous differential equation is $m^4 + 4\gamma^4 = 0$, with roots $m = \gamma(\pm 1 \pm i)$. The general solution of the differential equation is

$$u(x) = H + e^{\gamma x}(c_1 \cos(\gamma x) + c_2 \sin(\gamma x)) + e^{-\gamma x}(c_3 \cos(\gamma x) + c_4 \sin(\gamma x)).$$

Boundedness as $x \to \infty$ requires $c_1 = c_2 = 0$. Then the boundary conditions at $x = 0$ become

$$u(0) = 0: \quad H + c_3 = 0$$
$$u''(0): \quad 2\gamma^2 c_4 = 0.$$

Finally, $u(x) = H - He^{-\gamma x}\cos(\gamma x)$.

17. In each part of the wall (that is, in each subinterval) u is a straight line function.

$$u(x) = \begin{cases} c_1 + c_2 x & 0 < x < \alpha a \\ c_3 + c_4 x & \alpha a < x < a \end{cases}$$

The four conditions are:

$$u(0) = T_0: \qquad\qquad c_1 = T_0$$
$$u(a) = T_1: \qquad\qquad c_3 + c_4 a = T_1$$
$$u(\alpha a-) = u(\alpha a+): \qquad c_1 + c_2 \alpha a = c_3 + c_4 \alpha a$$
$$\kappa_1 u'(\alpha a-) = \kappa_2 u'(\alpha a+): \quad \kappa_1 c_2 = \kappa_2 c_4.$$

These can be solved by successive substitution. Check the answer by trying $\alpha = 0$ and $\alpha = 1$.

19. Rewrite the differential equation as $(e^x u')' = -e^{-x}$. Integrate: $e^x u' = e^{-x} + c_1$. Then $u' = e^{-2x} + c_1 e^{-x}$ and $u = -\frac{1}{2}e^{-2x} - c_1 e^{-x} + c_2$. Now apply the boundary conditions

$$u(0) = 0: \quad -\frac{1}{2} - c_1 + c_2 = 0$$

$$u(a) = 0: \quad -\frac{1}{2}e^{-2a} - c_1 e^{-a} + c_2 = 0$$

These can be solved by elimination to find

$$-\frac{1}{2}(1 - e^{-2a}) - c_1(1 - e^{-a}) = 0$$

or

$$c_1 = -\frac{(1 - e^{-2a})}{2(1 - e^{-a})} = -\frac{(1 + e^{-a})}{2}; \quad c_2 = \frac{1}{2} + c_1 = -\frac{e^{-a}}{2}.$$

The Answer is written in a form that makes it easy to check the boundary conditions.

21. The general solution of the differential equation and its derivative are:

$$u(x) = c_1 \cosh(px) + c_2 \sinh(px); \quad u'(x) = c_1 p \sinh(px) + c_2 p \cosh(px).$$

The following equations come from the boundary conditions.

a. $c_1 = 0$; $c_2 \sinh(pa) = 1$

b. $c_1 = 1$; $\cosh(pa) + c_2 \sinh(pa) = 0$

c. $c_2 p = 0$; $c_1 \cosh(pa) = 1$

d. $c_1 = 1$; $p \sinh(pa) + c_2 p \cosh(pa) = 0$. Use hyperbolic identities (see Mathematical References) to change the form.

e. $c_2 p = 1$; $c_1 p \sinh(pa) + \cosh(pa) = 0$

f. $c_2 p = 0$; $c_1 p \sinh(pa) = 1$

23. Let $u = xv$. The differential equation becomes

$$xv'' + 2v' - \frac{2x}{1 - x^2}(xv' + v) + \frac{2}{1 - x^2}xv = 0$$

or

$$xv'' + 2\left(\frac{1 - 2x^2}{1 - x^2}\right)v' = 0.$$

Now

$$\frac{v''}{v'} = -2\frac{1 - 2x^2}{x(1 - x^2)} = -2\left[\frac{1}{x} - \frac{\frac{1}{2}}{1 - x} + \frac{\frac{1}{2}}{1 + x}\right]$$

by partial fractions. Next, integrate to obtain

$$\ln v' = -2\left[\ln x + \frac{1}{2}\ln(1 - x) + \frac{1}{2}\ln(1 + x)\right] + c$$

$$v' = e^c\frac{1}{x^2(1 - x^2)} = e^c\left(\frac{1}{x^2} + \frac{1}{1 - x^2}\right) = e^c\left(\frac{1}{x^2} + \frac{\frac{1}{2}}{1 - x} + \frac{\frac{1}{2}}{1 + x}\right)$$

by partial fractions again. Finally

$$v = e^c\left(-\frac{1}{x} - \frac{1}{2}\ln(1 - x) + \frac{1}{2}\ln(1 + x)\right) + c_2$$

$$u = xv = e^c\left(\frac{x}{2}\ln\left(\frac{1 + x}{1 - x}\right) - 1\right) + c_2 x.$$

Of course, e^c is just a constant.

25. See Answers.

27. See Answers

29. The general solution (appropriate for the semi-infinite interval) is $u(x) = c_1 e^{ax} + c_2 e^{-ax}$. To obtain boundedness we must have $c_1 = 0$. The other boundary condition requires $c_2 = C_0$.

31. The characteristic equation of the homogeneous differential equation is $m^4 - \gamma^2 m^2 = 0$, with roots $m = 0, 0, \gamma, -\gamma$. The general solution of the homogeneous differential equation is

$$w_c(x) = c_1 + c_2 x + c_3 \cosh(\gamma x) + c_4 \sinh(\gamma x).$$

The inhomogeneity is constant, but constant and x are solutions of the homogeneous equation. Thus, our trial solution is $w_0 = A x^2$. Substituting in the differential equation shows that $A = -P/2\gamma^2$.

Now the general solution of the differential equation is

$$w(x) = -\frac{P}{\gamma^2}\frac{x^2}{2} + c_1 + c_2 x + c_3 \cosh(\gamma x) + c_4 \sinh(\gamma x).$$

We apply first the boundary conditions on w : $w(\pm\frac{1}{2}) = 0$. This becomes

$$-\frac{P}{\gamma^2}\left(\frac{1}{8}\right) + c_1 \pm c_2\frac{1}{2} + c_3 \cosh(\gamma/2) \pm c_4 \sinh(\gamma/2) = 0.$$

This is really two equations. The $\pm$ terms must be 0 and the remaining terms must also be 0:

$$c_2 \cdot \frac{1}{2} + c_4 \sinh(\gamma/2) = 0 \tag{1}$$

$$-\frac{P}{\gamma^2} \cdot \frac{1}{8} + c_1 + c_3 \cosh(\gamma/2) = 0. \tag{2}$$

The derivative of w is

$$w'(x) = -\frac{P}{\gamma^2}x + c_2 + c_3\gamma \sinh(\gamma x) + c_4\gamma \cosh(\gamma x).$$

Applying the boundary condition $w'(\pm\frac{1}{2}) = 0$ gives

$$-\frac{P}{\gamma^2}(\pm\frac{1}{2}) + c_2 \pm c_3\gamma \sinh(\gamma/2) + c_4\gamma \cosh(\gamma/2) = 0$$

Again, we get two equations:

$$-\frac{P}{\gamma^2} \cdot \frac{1}{2} + c_3\gamma \sinh(\gamma/2) = 0 \tag{3}$$

$$c_2 + c_4\gamma \cosh(\gamma/2) = 0 \tag{4}$$

Find c_3 directly from (3); then find c_1 from (2):

$$c_3 = \frac{P}{2\gamma^3 \sinh(\gamma/2)}, \quad c_1 = \frac{P}{8\gamma^2} - \frac{P}{2\gamma^3}\frac{\cosh(\gamma/2)}{\sinh(\gamma/2)}.$$

Equations (1) and (4) together force $c_2 = 0$, $c_4 = 0$.

Finally the solution of the boundary value problem is

$$w(x) = \frac{P}{2\gamma^2}\left[\frac{1}{4} - x^2 + \frac{\cosh(\gamma x) - \cosh(\gamma/2)}{\gamma \sinh(\gamma/2)}\right].$$

If P is positive, so is $w(x)$ for $-\frac{1}{2} < x < \frac{1}{2}$.

33. The solution of the problem follows the same lines as in #31:

$$w(x) = \frac{P}{2\lambda^2}\left[x^2 - \frac{1}{4} + \frac{\cos(\lambda x) - \cos(\lambda/2)}{\lambda \sin(\lambda/2)}\right].$$

However, the equivalents of Equations (1) and (4) are

$$c_2 \cdot \frac{1}{2} + c_4 \sin(\lambda/2) = 0$$

$$c_2 + c_4 \lambda \cos(\lambda/2) = 0$$

with determinant $\frac{1}{2}\lambda \cos(\lambda/2) - \sin(\lambda/2)$. This is 0 if $\tan(\lambda/2) = \lambda/2$. If this is true, c_2 and c_4 are undetermined: The shape is indefinite. This is a kind of buckling.

Chapter 1

1.1 Periodic Functions and Fourier Series

1.a. $f(x) = x, \quad -\pi < x < \pi; \quad f(x + 2\pi) = f(x)$

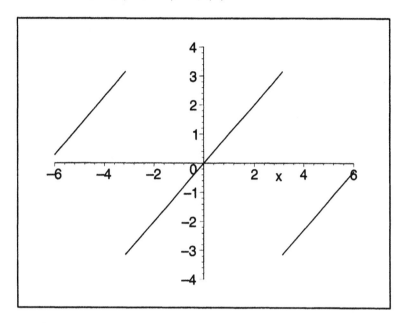

Exercise 1a

$$a_0 = 0; \quad a_n = 0;$$

$$b_n = \frac{1}{\pi} \int_{-\pi}^{\pi} \sin(nx)dx$$

$$= \frac{1}{\pi} \left[\frac{\sin(nx)}{n^2} - \frac{x\cos(nx)}{n} \right]_{-\pi}^{\pi}$$

$$= \frac{1}{\pi} \left[-\frac{\pi\cos(n\pi)}{n} + \frac{-\pi\cos(-n\pi)}{n} \right]$$

$$= \frac{-2\cos(n\pi)}{n} = \frac{2}{n}(-1)^{n+1}$$

(Recall: $\sin(n\pi) = 0$; $\cos(-n\pi) = \cos(n\pi)$,

$$\cos(n\pi) = 1(n = 0), \quad -1(n = 1), \quad 1(n = 2), \quad -1(n = 3), \cdots$$

so $-\cos(n\pi) = (-1)^{n+1}$. Don't try to extend this fact to non-integral values of n.)

b. $f(x) = |x|, \quad -\pi < x < \pi, \quad f(x + 2\pi) = f(x)$.

To do the integrations, you have to use the fact that $|x| = x$ if $x > 0$ but $|x| = -x$ if $x < 0$. Look at a graph for $-\pi < x < \pi$.

$a_0 = \frac{\pi}{2}$ by geometry; $b_n = 0$;

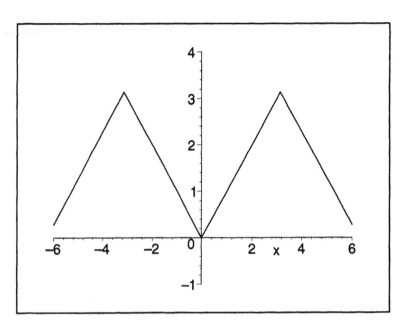

Exercise 1b

$$a_n = \frac{1}{\pi} \int_{-\pi}^{\pi} f(x)\cos(nx)dx = \frac{1}{\pi}\left[\int_{-\pi}^{0} -x\cos(nx)dx + \int_{0}^{\pi} x\cos(nx)dx\right]$$

c. $f(x) = \begin{cases} 0, & -\pi < x < 0 \\ 1, & 0 < x < \pi \end{cases}$ $\qquad f(x+2\pi) = f(x)$

$$a_0 = \frac{1}{2\pi} \int_{-\pi}^{\pi} f(x)dx = \frac{1}{2\pi} \int_{0}^{\pi} 1 \cdot dx = \frac{1}{2}$$

$$a_n = \frac{1}{\pi} \int_{-\pi}^{\pi} f(x)\cos(nx)dx = \frac{1}{\pi} \int_{0}^{\pi} \cos(nx)dx = \frac{\sin(nx)}{n\pi}\Big|_{0}^{\pi} = 0$$

$$b_n = \frac{1}{\pi} \int_{-\pi}^{\pi} f(x)\sin(nx)dx = \frac{1}{\pi} \int_{0}^{\pi} \sin(nx)dx = \frac{-\cos(nx)}{n\pi}\Big|_{0}^{\pi} = \frac{1-\cos(n\pi)}{n\pi}.$$

Note that $1 - \cos(n\pi) = 0$ if n is even or $= 2$ if n is odd.

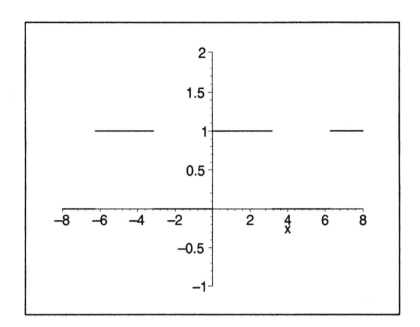

Exercise 1c

d.

$$f(x) = |\sin(x)| = \begin{cases} \sin(x), & 0 < x < \pi \\ -\sin(x), & -\pi < x < 0 \end{cases}$$

$$a_0 = \frac{1}{2\pi} \int_{-\pi}^{\pi} f(x)dx = \frac{1}{2\pi}\left[\int_{-\pi}^{0} -\sin(x)dx + \int_{0}^{\pi} \sin(x)dx\right]$$

$$= \frac{1}{2\pi}\left[\cos(x)\,\big|_{-\pi}^{0} - \cos(x)\,\big|_{0}^{\pi}\right]$$

$$= \frac{1}{2\pi}\left[1 - \cos(\pi) - (\cos(\pi) - 1)\right] = \frac{4}{2\pi} = \frac{2}{\pi}$$

$$a_n = \frac{1}{\pi} \int_{-\pi}^{\pi} f(x)\cos(nx)dx = \frac{1}{\pi}\left[\int_{-\pi}^{0} -\sin(x)\cos(nx)dx + \int_{0}^{\pi} \sin(x)\cos(nx)dx\right]$$

$$= \frac{1}{\pi}\left[-\frac{\cos((n-1)x)}{2(n-1)} + \frac{\cos((n+1)x)}{2(n+1)}\,\bigg|_{-\pi}^{0} + \frac{\cos((n-1)x)}{2(n-1)} - \frac{\cos((n+1)x)}{2(n+1)}\,\bigg|_{0}^{\pi}\right]$$

$$= \frac{1}{\pi}\left[\frac{-1+\cos((n-1)\pi)}{2(n-1)} + \frac{1-\cos((n+1)\pi)}{2(n+1)} + \frac{\cos((n-1)\pi)-1}{2(n-1)} - \frac{\cos((n+1)\pi)-1}{2(n+1)}\right]$$

$$= \frac{1}{\pi}\left[\frac{1-\cos((n+1)\pi)}{n+1} - \frac{1-\cos((n-1)\pi)}{n-1}\right].$$

Note that $\cos((n+1)\pi) = \cos((n-1)\pi)$ by periodicity, and both $= -\cos(n\pi)$

$$a_n = \frac{1+\cos(n\pi)}{\pi}\left[\frac{1}{n+1} - \frac{1}{n-1}\right] = \frac{-2}{\pi}\frac{1+\cos(n\pi)}{n^2-1}.$$

This formula doesn't work for $n = 1$, so compute $a_1 = 0$ separately.

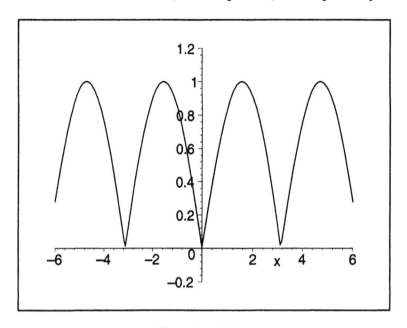

Exercise 1d

3. See the Answers

5. See the Answers

7. Use trigonometric identities.

a. $\cos^2(x) = \frac{1}{2}(1 + \cos(2x))$. This **is** a Fourier series: $a_0 = \frac{1}{2}$, $a_2 = \frac{1}{2}$, all b's and other a's are 0.

b. $\sin(x - \pi/6) = \sin(x)\cos(\pi/6) - \cos(x)\sin(\pi/6)$. This is a Fourier series with $a_1 = \sin(\pi/6) = \frac{1}{2}$, $b_1 = \cos(\pi/6) = \sqrt{3}/2$, all other coefficients are 0.

c. $\sin(x)\cos(2x) = \frac{1}{2}(\sin(-x) + \sin(3x))$; $b_1 = -\frac{1}{2}$, $b_3 = \frac{1}{2}$, all a's and all other b's are 0.

Chapter 1

1.2 Arbitrary Period and Half-Range Expansions

1.a. See Section 1.1, Exercise 1b. Change scale.

b. The periodic extension is an odd function, so $a_0 = 0$, $a_n = 0$

$$b_n = \frac{2}{2} \int_0^2 1 \cdot \sin\left(\frac{n\pi x}{2}\right) dx = \frac{2(1 - \cos(n\pi))}{n\pi}, \quad (a = 2).$$

c. The periodic extension is **even** and $2a = \frac{1}{2} - (-\frac{1}{2}) = 1$, so $a = \frac{1}{2}$

$$a_0 = \frac{1}{\frac{1}{2}} \int_0^{\frac{1}{2}} x^2 dx = \frac{1}{12}$$

$$a_n = \frac{2}{\frac{1}{2}} \int_0^{\frac{1}{2}} x^2 \cos\left(\frac{n\pi x}{\frac{1}{2}}\right) = 4 \int_0^{\frac{1}{2}} x^2 \cos(2n\pi x) dx$$

$$a_n = 4 \left[\frac{2x \cos(2n\pi x)}{(2n\pi)^2} + \frac{((2n\pi x)^2 - 2) \sin(2n\pi x)}{(2n\pi)^3} \right]\Big|_0^{\frac{1}{2}}$$

$$= 4 \left[\frac{\cos(n\pi)}{4n^2\pi^2} \right] = \frac{\cos(n\pi)}{n^2\pi^2}.$$

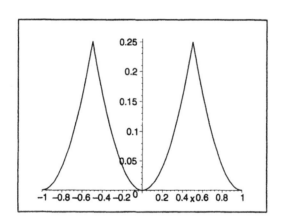

Exercise 1c

3. The formulas for $\bar{f}(x)$ the periodic extension with period 2a are: $\bar{f}(x) = f(x - 2ka)$, where k is the integer for which $k \leq \frac{x}{2a} < k + 1$. Then $x - 2ka$ is between 0 and 2a, so f is known (given) there. The sine coefficients of $\bar{f}$ are

$$b_n = \frac{1}{a} \int_{-a}^{a} \bar{f}(x) \sin\left(\frac{n\pi x}{a}\right) dx$$

$$= \frac{1}{a} \int_{0}^{2a} \bar{f}(x) \sin\left(\frac{n\pi x}{a}\right) dx$$

$$= \frac{1}{a} \int_{0}^{2a} f(x) \sin\left(\frac{n\pi x}{a}\right) dx.$$

The first step uses Exercise 5 of Section 1.1, and the second uses the fact that $\bar{f}(x) = f(x)$ for $0 < x < 2a$.

5. See the Answers.

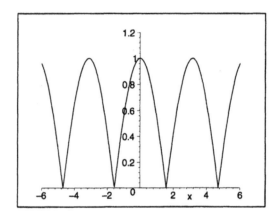

Exercise 5c

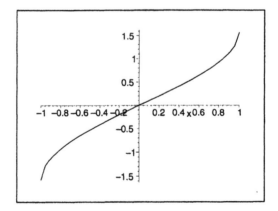

Exercise 5d

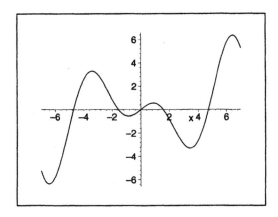

Exercise 5e

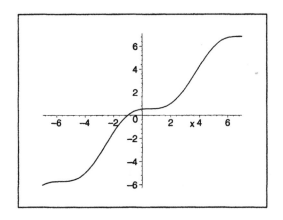

Exercise 5f

7.a. See the Example, Section 1.2.

b. The Fourier series is 1. That is, $a_0 = 1$, $a_n = 0$ for $n = 1, 2, \cdots$, $b_n = 0$ by evenness.

c. Look at the periodic extension. It is an odd function (by its nature, not because we made an odd extension). Therefore,

$$a_0 = a_n = 0, \quad 2a = \left(\frac{3}{2} - \left(-\frac{1}{2} \right) \right) \Rightarrow a = 1$$

$$b_n = \frac{2}{1} \int_0^1 f(x) \sin(n\pi x)$$

$$= 2 \left[\int_0^{\frac{1}{2}} x \sin(n\pi x) dx + \int_{\frac{1}{2}}^1 (1-x) \sin(n\pi x) dx \right]$$

$$= 2 \left[\left(\frac{\sin(n\pi x)}{(n\pi)^2} - \frac{x\cos(n\pi x)}{n\pi} \right) \Big|_0^{\frac{1}{2}} + \left(-\frac{(1-x)\cos(n\pi x)}{n\pi} - \frac{\sin(n\pi x)}{(n\pi)^2} \Big|_{\frac{1}{2}}^1 \right) \right]$$

$$= 2 \left[\frac{\sin(n\pi/2)}{(n\pi)^2} - \frac{\frac{1}{2}\cos(n\pi/2)}{n\pi} - \frac{-\frac{1}{2}\cos(n\pi/2)}{n\pi} - \frac{-\sin(n\pi/2)}{n\pi} \right]$$

$$= 4 \frac{\sin(n\pi/2)}{(n\pi)^2}$$

9. Look at the graph of $\sin(2\pi x/a)$ on the interval $0 < x < a$. It is symmetric in the point $(a/2, 0)$, just as $\sin(\alpha x)$ is symmetric in the origin. A function that is symmetric in the **line** $x = a/2$ will have $b_2 = 0$ and all b's with even subscript $= 0$. For example if $f(x) = 1$, $0 < x < a$ and $f(x)$ is odd, then $b_2 = b_4 = \cdots = 0$. Find another example.

11.a. Cosine series: $a_0 = 1$, $a_n = 0$ for $n \geq 1$. This function is among the functions in the cosine series. Sine series: $\sum b_n \sin(n\pi x/a)$ with

$$b_n = \frac{2}{a} \int_0^a 1 \cdot \sin\left(\frac{n\pi x}{a}\right) dx = \frac{-2\cos(n\pi x/a)}{n\pi} \Big|_0^a = \frac{2(1-\cos(n\pi))}{n\pi}.$$

b. See the example at the end of Section 1.2.

c. Notice that the interval is $0 < x < 1$, so the graph is just part of a sine arch.

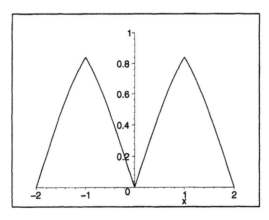

Exercise 11c even

Sine series $\sum_1^\infty b_n \sin(n\pi x)$ because $a = 1$ here; and

$$b_n = \frac{\sin(n\pi - 1)}{n\pi - 1} - \frac{\sin(n\pi + 1)}{n\pi + 1}.$$

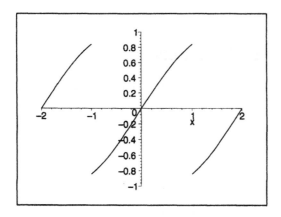

Exercise 11c odd

The numerators are the same except for sign and can be simplified. Cosine series $a_0 + \sum_1^\infty a_n \cos(n\pi x)$, $a_0 = 1 - \cos(1)$ and

$$a_n = \frac{\cos(n\pi - 1) - 1}{n\pi - 1} - \frac{\cos(n\pi + 1) - 1}{n\pi + 1}.$$

The numerators are the same and can be simplified.

d. The sine series is $\sum_1^\infty b_n \sin(nx)$ and $b_1 = 1$, $b_n = 0$ for $n > 1$. Of course! The cosine series is $a_0 + \sum_1^\infty a_n \cos(nx)$, and

$$a_n = \frac{1}{\pi}\left[\frac{\cos((n-1)\pi) - 1}{n - 1} - \frac{\cos((n+1)\pi) - 1}{n + 1}\right], \quad (n \neq 1).$$

The numerators are equal (why?) and their mutual value is $-1 - \cos(n\pi)$.

For a_1 you have to do a separate calculation, because a denominator is 0 otherwise. It turns out that $a_1 = 0$.

13. The even periodic extension of a function continuous on $0 < x < a$ is continuous everywhere, because $\bar{f}_e(0-) = f(0+)$ and $\bar{f}_e(-a+) = f(a-)$. For the odd periodic extension, however, $\bar{f}_0(0-) = -f(0+)$ and $\bar{f}_0(-a+) = -f(a-)$, so continuity of $\bar{f}_0$ fails at $2na$ unless $f(0+) = 0$ and at $(2n - 1)a$ unless $f(a-) = 0$.

15.a. $|h(-x)| = |-h(x)| = |h(x)|$

b. $f(|-x|) = f(|x|)$

c. $f(g(-x)) = f(g(x))$

d. $g(h(-x)) = g(-h(x)) = g(h(x))$.

Chapter 1

1.3 Convergence of Fourier Series

1.a. The function is continuous with corners (discontinuities in the derivative) at 0 and 1. Sectionally smooth.

b. The function is continuous but fails to be sectionally smooth because of a vertical tangent at $x = 0$.

c. Since $\cos(x/2) \geq 0$ for $-\pi < x < \pi$, the function is well defined (and real) and continuous. However, there are vertical asymptotes at $x = \pm\pi$ which make it fail to be sectionally smooth.

d. This function is continuous and has continuous derivatives of all orders. Sectionally smooth.

e. This function has a vertical asymptote at $x = \pi$, so it is not sectionally smooth on the interval $0 < x < \pi$.

3. If f is continuous, sectionally smooth and periodic, its Fourier series converges to $f(x)$ at every x.

5.a.

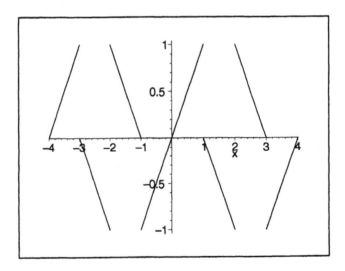

Figure 1: Exercise 5a

b. If the graph is cut into vertical panels with cut lines at $x = \pm1$, ±2, ±3, ±4 then the graph is smooth (straight-line segment) on each panel.

c. At $x = 1$: $+\frac{1}{2}$. At $x = 2$: 0. At $x = 9.6$, same as at 1.6 because the period is 4. Therefore the series converges there to -0.6. At -3.8, same as at $+0.2$, the series converges to 0.2.

7. The series converges to an even function, so $B = 0$. The coefficient a_0 is missing; it must be 0. Therefore

$$0 = \frac{1}{\pi} \int_0^\pi (A + Cx^2)dx = A + C\frac{\pi^2}{3}.$$

Also, $a_n = (-1)^n/n^2$ for $n \geq 1$, so

$$\frac{(-1)^n}{n^2} = \frac{2}{\pi} \int_0^\pi (A + Cx^2)\cos(nx)dx = \frac{4\cos(n\pi)}{n^2} \cdot C.$$

Therefore, $C = 1/4$, $A = -\pi^2/12$.

9.a. $f(x) = \sqrt{1 - x^2}$, $-1 < x < 1$

b. $a_0 = \frac{1}{2}$ area under curve $= \frac{\pi}{4}$.

c. $f(x)$ fails to be sectionally smooth because of vertical tangents at ± 1.

d. The theorem tells us nothing because $f(x)$ is not sectionally smooth.

Chapter 1

1.4 Uniform Convergence

1.a. The periodic function is not continuous at ± 1 etc., so convergence cannot be uniform.

b. Like a, the periodic function has jumps at $\pm \pi$ etc.

c. This function is continuous and has continuous derivatives of all orders. "Convergence" is uniform – it's hardly reasonable to speak of convergence, since the series has only one term.

d. This periodic function is continuous, and its derivative is continuous except for jumps at $\pm \pi$ etc. Convergence is uniform.

e. The periodic function has jumps, so convergence cannot be uniform.

f. The periodic function and its (periodic) derivative are both continuous for all x. Convergence is uniform.

g. Same answer as for part f. Note the close relationship between the functions.

3. Sine series: b. is a finite series; for a,c,d,e,f the odd periodic extension has jumps, so the convergence cannot be uniform. Cosine series: all of the even periodic extensions are continuous with sectionally continuous derivatives, so the convergence is uniform.

5. For each one, we must test whether $* \sum(|a_n| + |b_n|)$ converges.

It is useful to know these series from calculus: $\sum \frac{1}{n}$ does not converge, $\sum \frac{1}{n^2}$ does converge. Also if $|c_n| \geq K/n$, $\sum |c_n|$ does not converge; if $|d_n| \leq K/n^2$, $\sum |d_n|$ does converge.

a. $|a_n| + |b_n| \leq \frac{1/\pi^2}{n^2}$, so $*$ converges, and convergence of the Fourier series is uniform.

b. If n is odd, $b_n = \frac{2/\pi}{n}$, so the series $\sum |b_n|$ does not converge.

c. $|a_n| + |b_n| \leq \frac{8/3}{n^2}$, so $*$ converges and the Fourier series converges uniformly.

d. $|a_n| + |b_n| \leq 2e^{-n\pi/2}$, so $*$ converges and the Fourier series converges uniformly.

Chapter 1

1.5 Operations on Fourier Series

1. See Answers.

3. $f'(x) = 1$, $0 < x < \pi$. The Fourier sine series cannot be differentiated term by term because the odd periodic extension of f is not continuous. The cosine series can be differentiated, to produce the sine series for the square-wave function.

5. For the sine series: $f(0+) = 0$ and $f(a-) = 0$ are necessary to ensure that the odd periodic extension is continuous. Then the differentiated series converges to $f'(x)$ wherever $f''(x)$ exists, per Theorem 6. For the cosine series: no extra conditions are needed.

7. The function $\ln(|2\cos(x/2)|)$ has vertical asymptotes at odd multiples of π. It is not continuous, and its series cannot be differentiated term by term.

9.a. The coefficients b_n cannot get large as $n \to \infty$ (details in Section 1.6), and $e^{-n^2 t} \to 0$ so quickly that term-by-term differentiation is okay by Theorem 7.

b. At $x = 0$ and $x = \pi$, every term of the series is 0.

c. This follows from the convergence theorem in Section 1.3.

Chapter 1

1.6 Mean Error and Convergence in Mean

1. Use Exercise 5 of Section 1.5 to identify the Fourier series of $f(x) = \ln(|2\cos(x/2)|)$. Then use item 2 of the summary to identify the required integral as being equal to $\sum_{n=1}^{\infty} \frac{1}{n^2}$. Finally, from Exercise 1 of Section 1.5, this series has sum $\pi^2/6$.

3. Check the Summary.

a. $\int_{-1}^{1} |x| dx = 1$, and the function is even, so $b_n = 0$ and $\sum_{1}^{\infty} a_n^2$ converges.

b. $\int_{-1}^{1} \frac{1}{|x|} dx$ does not converge. The integrals for a_n and b_n, although improper, do converge. However the series $\sum_{n=1}^{\infty} (a_n^2 + b_n^2)$ does not converge.

5. The integral must be improper and cannot converge. If the integral had a finite value then $\sum_{1}^{\infty} (a_n^2 + b_n^2)$ would also have a finite value. But $\sum b_n^2 = \sum 1/n$ is known to be divergent.

Chapter 1

1.7 Proof of Convergence

1. See Answers.

3. Since we are using $x = 0$ (where $f(x)$ has a corner and $f(0) = 0$) the function is

$$\phi(y) = \frac{|y|}{2\sin(y/2)} \cos(y/2).$$

For $y > 0$, the numerator is just y, and $\phi(0+) = 1$. Vice-versa, for $y < 0$, the numerator is $-y$, and $\phi(0-) = -1$.

5. See Answers.

Chapter 1

1.8 Numerical Fourier Coefficients

1. Add a column to Table 2 and fill with the values of $\cos(6x_i)$. The values are all 1 or –1. Use the last equation of Eq.(7) with $s = 6$ to find $\hat{a}_6$. The table below, produced in Excel, shows the information for calculating all seven coefficients.

n:	0	1	2	3	4	5	6	sin(x)/x
i	xi	cos(0*xi)	cos(1*xi)	cos(2*xi)	cos(3*xi)	cos(4*xi)	cos(5*xi)	cos(6*xi)	
0	0	1	1	1	1	1	1	1	1
1	0.524	1	0.866	0.5	0	-0.5	-0.866	-1	0.955
2	1.047	1	0.5	-0.5	-1	-0.5	0.5	1	0.827
3	1.571	1	0	-1	0	1	0	-1	0.637
4	2.094	1	-0.5	-0.5	1	-0.5	-0.5	1	0.413
5	2.618	1	-0.866	0.5	0	-0.5	0.866	-1	0.191
6	3.142	1	-1	1	-1	1	-1	1	0
	an:	0.587	0.456	-0.061	0.029	-0.019	0.015	-0.007	

3. The tables below show the cosine and sine values needed to compute the approximate coefficients. Following is a graph of the sum of the series using the approximate coefficents. The curve produced by that series fits the data exactly and interpolates between data points.

n:	0	1	2	3	4	5	6		Data
k	k*pi/6								
0	0.0	1.0	1.0	1.0	1.0	1.0	1.0	1.0	0.75
1	0.524	1.000	0.866	0.500	0.000	-0.500	-0.866	-1.000	0.6
2	1.047	1.000	0.500	-0.500	-1.000	-0.500	0.500	1.000	0.65
3	1.571	1.000	0.000	-1.000	0.000	1.000	0.000	-1.000	1.15
4	2.094	1.000	-0.500	-0.500	1.000	-0.500	-0.500	1.000	1.8
5	2.618	1.000	-0.866	0.500	0.000	-0.500	0.866	-1.000	2.25
6	3.142	1.000	-1.000	1.000	-1.000	1.000	-1.000	1.000	2.35
7	3.665	1.000	-0.866	0.500	0.000	-0.500	0.866	-1.000	2.15
8	4.189	1.000	-0.500	-0.500	1.000	-0.500	-0.500	1.000	1.75
9	4.712	1.000	0.000	-1.000	0.000	1.000	0.000	-1.000	1.05
10	5.236	1.000	0.500	-0.500	-1.000	-0.500	0.500	1.000	1
11	5.760	1.000	0.866	0.500	0.000	-0.500	-0.866	-1.000	0.9
	an:	1.3667	-0.844	0.2083	0.05	-0.042	-0.006	0.0167	

	n:	0	1	2	3	4	5	6		Data
k	k*pi/6									
0	0.0	0.0	0.0	0.0	0.0	0.0	0.0	0.0		0.75
1	0.524	0.000	0.500	0.866	1.000	0.866	0.500	0.000		0.6
2	1.047	0.000	0.866	0.866	0.000	-0.866	-0.866	0.000		0.65
3	1.571	0.000	1.000	0.000	-1.000	0.000	1.000	0.000		1.15
4	2.094	0.000	0.866	-0.866	0.000	0.866	-0.866	0.000		1.8
5	2.618	0.000	0.500	-0.866	1.000	-0.866	0.500	0.000		2.25
6	3.142	0.000	0.000	0.000	0.000	0.000	0.000	0.000		2.35
7	3.665	0.000	-0.500	0.866	-1.000	0.866	-0.500	0.000		2.15
8	4.189	0.000	-0.866	0.866	0.000	-0.866	0.866	0.000		1.75
9	4.712	0.000	-1.000	0.000	1.000	0.000	-1.000	0.000		1.05
10	5.236	0.000	-0.866	-0.866	0.000	0.866	0.866	0.000		1
11	5.760	0.000	-0.500	-0.866	-1.000	-0.866	-0.500	0.000		0.9
	bn:	0	-0.043	-0.115	-0.05	4E-16	0.0433	3E-16		

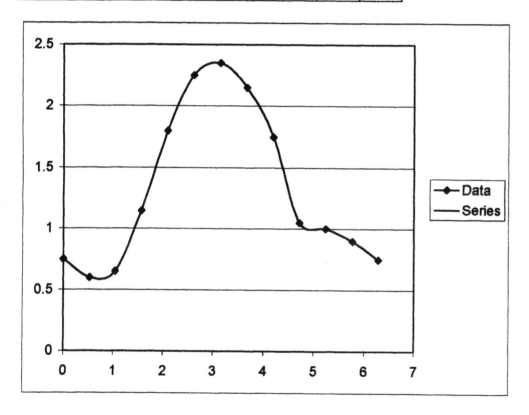

Chapter 1

1.9 Fourier Integral

1.a. For the even extension,

$$
\begin{aligned}
A(\lambda) &= \frac{2}{\pi} \int_0^\infty e^{-x} \cos(\lambda x) dx \\
&= \frac{2}{\pi} \left. \frac{e^{-x}(-\cos(\lambda x) + \lambda \sin(\lambda x))}{1 + \lambda^2} \right|_0^\infty \\
&= \frac{2}{\pi} \frac{2}{\pi} \frac{1}{1 + \lambda^2}.
\end{aligned}
$$

For the odd extension,

$$
\begin{aligned}
B(\lambda) &= \frac{2}{\pi} \int_0^\infty e^{-x} \sin(\lambda x) dx \\
&= \frac{2}{\pi} \left. \frac{e^{-x}(-\sin(\lambda x) - \lambda \cos(\lambda x))}{1 + \lambda^2} \right|_0^\infty \\
&= \frac{2}{\pi} \frac{\lambda}{1 + \lambda^2}.
\end{aligned}
$$

b. Note the limits, which come from the function definition.

$$
A(\lambda) = \frac{2}{\pi} \int_0^1 \cos(\lambda x) dx = \frac{2 \sin(\lambda)}{\pi \lambda}
$$

$$
B(\lambda) = \frac{2}{\pi} \int_0^1 \sin(\lambda x) dx = \frac{2(1 - \cos(\lambda))}{\pi \lambda}
$$

c. Note the limits.

$$
A(\lambda) = \frac{2}{\pi} \int_0^\pi (\pi - x) \cos(\lambda x) dx
$$

$$
= \frac{2}{\pi} \left[(\pi - x) \left. \frac{\sin(\lambda x)}{\lambda} \right|_0^\pi + \int_0^\pi \frac{\sin(\lambda x)}{\lambda} dx \right]
$$

$$
= \frac{2}{\pi} \left[\frac{1 - \cos(\lambda \pi)}{\lambda^2} \right].
$$

d.

$$B(\lambda) = \frac{2}{\pi} \int_0^\pi (\pi - x) \sin(\lambda x)\, dx$$

$$= \frac{2}{\pi} \left[(\pi - x) \frac{-\cos(\lambda x)}{\lambda} \Big|_0^\pi - \int_0^\pi \frac{\cos(\lambda x)}{\lambda}\, dx \right]$$

$$= \frac{2}{\pi} \left[\frac{\pi}{\lambda} - \frac{\sin(\lambda x)}{\lambda^2} \Big|_0^\pi \right] = \frac{2}{\pi} \left[\frac{\pi}{\lambda} - \frac{\sin(\lambda \pi)}{\lambda^2} \right]$$

$$= \frac{2}{\lambda} - \frac{2\sin(\lambda \pi)}{\pi \lambda^2}.$$

3.a. We need to find

$$A(\lambda) = \frac{2}{\pi} \int_0^\infty \frac{1}{1 + x^2} \cos(\lambda x)\, dx.$$

The integral cannot be found in the usual way. However, we know (see Exercise 1a. or Example 3)

$$\frac{2}{\pi} \int_0^\infty \frac{\cos(\lambda x)}{1 + \lambda^2}\, d\lambda = e^{-|x|}, \quad -\infty < x < \infty.$$

Therefore

$$\frac{2}{\pi} \int_0^\infty \frac{1}{1 + x^2} \cos(\lambda x)\, dx = e^{-|\lambda|}$$

and (since the function is even, sectionally smooth and continuous)

$$\frac{1}{1 + x^2} = \int_0^\infty e^{-\lambda} \cos(\lambda x)\, d\lambda.$$

This equality can be confirmed by integrating directly.

b. The given function is even. We need to find

$$A(\lambda) = \frac{2}{\pi} \int_0^\infty \frac{\sin(x)}{x} \cos(\lambda x)\, dx.$$

Again, consulting Example 2, we find that

$$\frac{2}{\pi} \int_0^\infty \frac{\sin(\lambda)}{\lambda} \cos(\lambda x)\, d\lambda = \begin{cases} 1, & |x| < 1 \\ \\ 0, & |x| > 1 \end{cases}.$$

Therefore, by exchanging x and λ, we have

$$A(\lambda) = \begin{cases} 1, & 0 < \lambda < 1 \\ \\ 0, & 1 < \lambda \end{cases}$$

and

$$\frac{\sin(x)}{x} = \int_0^1 \cos(\lambda x)d\lambda,$$

which can be confirmed directly.

5.a. See Example.

b. From the same Example,

$$A(\lambda) = -\frac{1 + \cos(\lambda \pi)}{\pi(\lambda^2 - 1)}, \quad B(\lambda) = -\frac{\sin(\lambda \pi)}{\pi(\lambda^2 - 1)}$$

$$f(x) = -\frac{1}{\pi}\int_0^\infty \frac{(1 + \cos(\lambda \pi))\cos(\lambda x) + \sin(\lambda \pi)\sin(\lambda x)}{\lambda^2 - 1}d\lambda.$$

c. This function is even. The Example again gives

$$A(\lambda) = -\frac{2(1 + \cos(\lambda \pi))}{\pi(\lambda^2 - 1)}.$$

7. In 6c, change variable from x to λ with $x = \lambda z$, $dx = zd\lambda$. Then

$$\int_0^\infty \frac{\sin(x)}{x}dx = \int_0^\infty \frac{\sin(\lambda z)}{\lambda z}zd\lambda = \int_0^\infty \frac{\sin(\lambda z)}{\lambda}d\lambda$$

if $z > 0$. If $z < 0$ the sign changes. If $z = 0$ the change of variable is not legitimate, but the integrand is 0, giving 0 value for the integral.

Chapter 1

1.10 Complex Methods

1.

$$a_n - ib_n = \frac{1}{\pi} \int_{-\pi}^{\pi} e^{\alpha x} e^{-inx} dx$$

$$\overset{[1]}{=} \frac{1}{\pi} \int_{-\pi}^{\pi} e^{(\alpha-in)x} dx = \frac{1}{\pi} \frac{e^{(\alpha-in)x}}{\alpha - in} \Bigg|_{-\pi}^{\pi}$$

$$= \frac{1}{\pi} \frac{e^{\alpha\pi} e^{-in\pi} - e^{-\alpha\pi} e^{+in\pi}}{\alpha - in}$$

$$\overset{[2]}{=} \frac{1}{\pi} \frac{(e^{\alpha\pi} - e^{-\alpha\pi}) \cos(n\pi)}{\alpha - in}$$

$$= \frac{1}{\pi} \frac{(e^{\alpha\pi} - e^{-\alpha\pi}) \cos(n\pi)}{\alpha^2 + n^2} (\alpha + in)$$

$$a_n \overset{[3]}{=} \frac{2}{\pi} \sinh(\alpha\pi) \frac{\alpha \cos(n\pi)}{\alpha^2 + n^2}$$

$$b_n = \frac{-2}{\pi} \sinh(\alpha\pi) \frac{n \cos(n\pi)}{\alpha^2 + n^2}.$$

[1] Use properties of exponents.

[2] $\sin(n\pi) = 0$.

[3] Use definition of sinh.

3. a. $C(\lambda) = \frac{1}{2\pi} \int_0^{\infty} e^{-x} e^{-i\lambda x} dx$

$$= \frac{1}{2\pi} \frac{e^{-(1+i\lambda)x}}{-(1+i\lambda)} \Bigg|_0^{\infty} = \frac{1}{2\pi} \frac{1}{1+i\lambda}$$

b. $C(\lambda) = \frac{1}{2\pi} \int_0^{\pi} \sin(x) e^{-i\lambda x} dx$

$$= \frac{1}{2\pi} \int_0^{\pi} \frac{e^{ix} - e^{-ix}}{2i} e^{-i\lambda x} dx$$

$$= \frac{1}{4\pi i} \left[\frac{e^{i(1-\lambda)x}}{i(1-\lambda)} - \frac{e^{-i(1+\lambda)x}}{-i(1+\lambda)} \right] \Bigg|_0^{\pi}$$

$$= \frac{1}{4\pi i} \left[\frac{e^{i(1-\lambda)\pi} - 1}{i(1-\lambda)} + \frac{e^{-i(1+\lambda)\pi} - 1}{i(1+\lambda)} \right]$$

$$\overset{[1]}{=} \frac{-(1+e^{-i\lambda\pi})}{-4\pi}\left(\frac{1}{1-\lambda}+\frac{1}{1+\lambda}\right)$$

$$= \frac{1+e^{-i\lambda\pi}}{2\pi(1-\lambda^2)}$$

[1]: $e^{i\pi} = e^{-i\pi} = -1$ so both numerators are $-e^{-i\lambda\pi} - 1$.

5. See Answers.

7. a. $f(x) = \displaystyle\int_{-\infty}^{\infty} C(\lambda)e^{i\lambda x}d\lambda = \int_{-1}^{1} e^{i\lambda x}d\lambda$

$$= \frac{e^{i\lambda x}}{ix}\bigg|_{-1}^{1} = \frac{e^{ix}-e^{-ix}}{ix} = \frac{2\sin(x)}{x}$$

b. $f(x) = \displaystyle\int_{-\infty}^{\infty} e^{-|\lambda|}e^{i\lambda x}d\lambda$. Break the integration at 0.

$$\int_{0}^{\infty} e^{-\lambda}e^{i\lambda x}d\lambda = \frac{e^{-\lambda+i\lambda x}}{-1+ix}\bigg|_{0}^{\infty} = \frac{1}{1-ix}$$

$$\int_{-\infty}^{0} e^{\lambda}e^{i\lambda x}d\lambda = \frac{e^{\lambda+i\lambda x}}{1+ix}\bigg|_{-\infty}^{0} = \frac{1}{1+ix}$$

$$f(x) = \frac{1}{1-ix}+\frac{1}{1+ix} = \frac{2}{1+x^2}$$

Chapter 1

1.11 Applications of Fourier Series and Integrals

1. The function $r(t)$ has period 4π, so

$$r(t) = a_0 + \sum_{n=1}^{\infty} a_n \cos(nt/2) + b_n \sin(nt/2).$$

$$a_0 = \frac{1}{2}, \quad a_n = 0, \quad b_n = -\frac{1}{n\pi}$$

$$r(t) = \frac{1}{2} - \sum_{n=1}^{\infty} \frac{1}{n\pi} \sin(nt/2).$$

Now assume that the particular solution u has the form

$$u(t) = A_0 + \sum_{n=1}^{\infty} A_n \cos(nt/2) + B_n \sin(nt/2).$$

Following the example in Part A, we have

$$A_0 = \frac{1}{2 \times 1.04}$$

$$A_n = \frac{-0.4n(-1/n\pi)}{(1.04 - n^2)^2 + (0.4n)^2} = \frac{0.4/\pi}{(1.04 - n^2)^2 + (0.4n)^2}$$

$$B_n = \frac{(1.04 - n^2)(-1/n\pi)}{(1.04 - n^2)^2 + (0.4n)^2} = -\frac{1}{n\pi} \frac{1.04 - n^2}{(1.04 - n^2)^2 + (0.4n)^2}$$

3. Let $T/EI = \gamma^2$, $w/EI = K$. The function $h(x)$ has this sine series:

$$h(x) = \sum_{n=1}^{\infty} \frac{8\sin(n\pi/2)}{n^2\pi^2} \sin(n\pi x/L).$$

The sine series is chosen because all of the $\sin(n\pi x/L)$ are 0 at $x = 0$ and at $x = L$.

Following the example of Part B, assume that $u(x)$ also has a sine series

$$u(x) = \sum_{n=1}^{\infty} B_n \sin(n\pi x/L).$$

The boundary conditions are satisfied.

$$u''(x) = \sum_{n=1}^{\infty} -B_n(n\pi/L)^2 \sin(n\pi x/L).$$

The differential equation becomes

$$\sum_{n=1}^{\infty} B_n(-(n\pi/L)^2 - \gamma^2) \sin(n\pi x/L) = K \sum_{n=1}^{\infty} \frac{-8\sin(n\pi/2)}{n^2\pi^2} \sin(n\pi x/L).$$

By matching coefficients, we find that

$$-B_n((n\pi/L)^2 + \gamma^2) = \frac{8K\sin(n\pi/2)}{n^2\pi^2}$$

$$B_n = \frac{8K\sin(n\pi/2)}{((n\pi/L)^2 + \gamma^2)n^2\pi^2},$$

and this gives $u(x)$.

5. See the Figures below for $\Omega = 4$, $N = 2$:

$$e^{-t^2} \cong \sum_{n=-2}^{2} e^{-(n\pi/4)^2}\frac{\sin(4t - n\pi)}{4t - n\pi}.$$

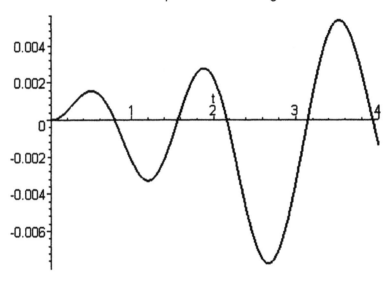

Difference between partial sum and target function

Figure 1: Exercise 5a

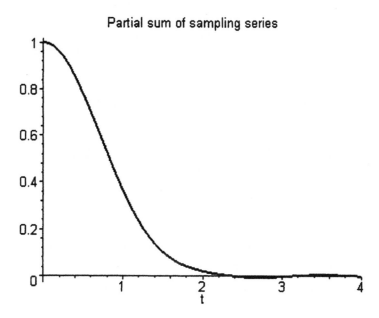

Figure 2: Exercise 5b

Chapter 1

Miscellaneous Exercises

1.

$$b_n = \frac{2}{\pi}\left[\int_0^\alpha \frac{x}{\alpha}\sin(nx)\,dx + \int_\alpha^{\pi-\alpha}\sin(nx)\,dx + \int_{\pi-\alpha}^\pi \frac{\pi-x}{\alpha}\sin(nx)\,dx\right]$$

First integral: $\dfrac{2}{\alpha\pi}\left[\dfrac{\sin(n\alpha)}{n^2} - \dfrac{\alpha\cos(n\alpha)}{n}\right]$

Second: $\dfrac{2}{\pi}\left[\dfrac{\cos(n\alpha) - \cos(n(\pi-\alpha))}{n}\right] = \dfrac{2}{\pi}\dfrac{\cos(n\alpha)}{n}(1-(-1)^n)$

Third: $(1-(-1)^n)$ times the first. To see this, change variables: $y = \pi - x$.
$b_n = \dfrac{2}{\pi}\dfrac{\sin(n\alpha)}{n^2\alpha}(1-(-1)^n)$. This is equivalent to the Answer.

3. As $\alpha \to 0$, $\sin(n\alpha)/n\alpha \to 1$, so $b_n \to \dfrac{2(1-(-1)^n)}{n\pi}$, the sine coefficient of the odd square wave.

5. Note that the function is continuous. Use integration by parts.

$$b_n = \frac{2}{a}\left[\int_0^{\alpha a}\frac{hx}{\alpha a}\sin\left(\frac{n\pi x}{a}\right)dx + \int_{\alpha a}^a \frac{h(a-x)}{(1-\alpha)a}\sin\left(\frac{n\pi x}{a}\right)dx\right]$$

$$= \frac{2h}{a^2}\left[\frac{1}{\alpha}\left(\frac{\sin(n\pi x/a)}{(n\pi/a)^2} - \frac{x\cos(n\pi x/a)}{n\pi/a}\right)\Big|_0^{\alpha a} + \frac{1}{1-\alpha}\left(\frac{-(a-x)\cos(n\pi x/a)}{n\pi/a} - \frac{\sin(n\pi x/a)}{(n\pi/a)^2}\right)\right]$$

$$= 2h\left[\frac{1}{\alpha}\left(\frac{\sin(n\pi\alpha)}{(n\pi)^2} - \frac{\alpha\cos(n\pi\alpha)}{n\pi}\right) + \frac{1}{1-\alpha}\left(\frac{(1-\alpha)\cos(n\pi\alpha)}{n\pi} + \frac{\sin(n\pi\alpha)}{(n\pi)^2}\right)\right]$$

$$= 2h\frac{\sin(n\pi\alpha)}{(n\pi)^2}\left(\frac{1}{\alpha} + \frac{1}{1-\alpha}\right).$$

7.a. $f_e(x) = 1$, $-a < x < a$. F.S.: 1.

b. $f_0(x) = \begin{cases} 1, & 0 < x < a \\ -1, & -a < x < 0 \end{cases}$. F.S.: $\displaystyle\sum_1^\infty \frac{2(1-(-1)^n)}{n\pi}\sin\left(\frac{n\pi x}{a}\right)$.

c. $\bar{f}(x) = 1$, $-\infty < x < \infty$. F.S.: 1

d. $\bar{f}_e(x) = 1$, $\infty < x < \infty$. F.S.: 1

e. $\bar{f}_0(x)$ is the odd square wave. F.S.: as in b.

f. Now the extended function is

$$\begin{cases} 1, & 0 < x < a \\ 0, & -a < x < 0 \end{cases}$$

$$a_0 = \frac{1}{2}$$

by geometry; $a_n = 0$ because the extended function is $\frac{1}{2}$ plus an odd square wave of height $\frac{1}{2}$; then $b_n = (1 - \cos(n\pi))/n\pi$.

9. Note that the given function is the sum of the odd and even extensions of $f(x) = x$, $0 < x < a$. Use this fact to find $a_0 = a/2$, $a_n = 2a(\cos(n\pi) - 1)/(n\pi)^2$; $b_n = -2a\cos(n\pi)/(n\pi)$. The periodic function has discontinuities at $x = \pm a, \pm 3a$, etc. The table shows the sum of the series, determined by using the convergence theorem.

x	$-a$	$-a/2$	0	a	$2a$
sum	a	0	0	a	0

11. $a_0 = \frac{3}{4}$ by geometry.

$$a_n = \frac{2}{\pi}\left[\int_0^{\pi/2}\cos(nx)dx + \int_{\pi/2}^{\pi}\frac{1}{2}\cos(nx)dx\right] = \frac{2}{\pi}\left[\left.\frac{\sin(nx)}{n}\right|_0^{\pi/2} + \left.\frac{\sin(nx)}{2n}\right|_{\pi/2}^{\pi}\right] = \frac{\sin(\frac{n\pi}{2})}{n\pi}.$$

This function has discontinuities at $x = \pm\pi/2, \pm 3\pi/2$, etc. The convergence theorem gives the values in the table

x	0	$\pi/2$	π	$3\pi/2$	2π
sum	1	$3/4$	$1/2$	$3/4$	1

13. The odd periodic extension, period 2, accidentally has period 1 because of odd symmetry about the point $x = 1/2$. Discontinuities occur at $x = 0, \pm 1, \pm 2$, etc., and the sum of the series is 0 at those points.

15. The function is odd, continuous, and sectionally smooth, so $f(x) = \sum b_n \sin(nx)$ and

$$b_n = \frac{2}{\pi}\int_0^{\pi/2}\sin(2x)\sin(nx)dx$$

$$= \frac{2}{\pi}\left[\frac{\sin((n-2)x)}{2(n-2)} - \frac{\sin((n+2)x)}{2(n+2)}\right]_0^{\pi/2} = -\frac{\sin(n\pi/2)}{\pi}\left[\frac{1}{n-2} - \frac{1}{n+2}\right], (n \neq 2).$$

Note that $\sin((n \pm 2)\pi/2) = \sin(n\pi/2 \pm \pi) = -\sin(n\pi/2)$. The integration formula would require division by 0 if $n = 2$, so b_2 must be found separately.

17. Use $\cos(nx) = \text{Re}(e^{inx})$, then find the real part of $\sum_1^N e^{inx}$:

$$S = \sum_1^N e^{inx} = \frac{e^{i(N+1)x} - e^{ix}}{e^{ix} - 1}.$$

This can be reduced by standard algebra, but it is quicker to multiply numerator and denominator by $e^{-ix/2}$; so the sum becomes

$$S = \frac{e^{i(N+1/2)x} - e^{ix/2}}{e^{ix/2} - e^{-ix/2}}.$$

The denominator is $2i \sin(x/2)$:

$$S = \frac{\cos((N + 1/2)x) + i \sin((N + 1/2)x) - (\cos(x/2) + i \sin(x/2))}{2i \sin(x/2)}$$

$$\text{Re}(S) = \frac{\sin((N + 1/2)x) - \sin(x/2)}{2 \sin(x/2)}.$$

To continue the transformation, use the identity for a difference of sines (see Mathematical References).

19. $f(x) = \sum b_n \sin(nx)$, and

$$b_n = \frac{2}{\pi} \int_0^a \sin\left(\frac{\pi x}{a}\right) \sin(nx)dx$$

$$= \frac{2}{\pi} \left[\frac{\sin(nx - \pi x/a)}{2(n - \pi/a)} - \frac{\sin(nx + \pi x/a)}{2(n + \pi/a)} \right]\Bigg|_0^a$$

$$= \frac{1}{\pi} \left[\frac{\sin(na - \pi)}{n - \pi/a} - \frac{\sin(na + \pi)}{n + \pi/a} \right]$$

$$= -\frac{\sin(na)}{\pi} \left[\frac{1}{n - \pi/a} - \frac{1}{n + \pi/a} \right] = -\frac{\sin(na)}{\pi} \cdot \frac{2n}{n^2 - (\pi/a)^2}.$$

Note that $\sin(na \pm \pi)$ has the same value for either sign because of periodicity.

21. $A(\lambda) = \frac{1}{\pi} \int_0^a \cos(\lambda x)dx = \frac{\sin(\lambda a)}{\pi \lambda}$, $B(\lambda) = \frac{1}{\pi} \int_0^a \sin(\lambda x)dx = \frac{1 - \cos(\lambda a)}{\pi \lambda}$.

23. $B(\lambda) = \frac{2}{\pi} \int_0^\pi \sin(x) \sin(\lambda x)dx$

$$= \frac{2}{\pi} \left[\frac{\sin((\lambda - 1)x)}{2(\lambda - 1)} - \frac{\sin((\lambda + 1)x)}{2(\lambda + 1)} \right]\Bigg|_0^\pi$$

$$= \frac{1}{\pi} \left[\frac{\sin(\lambda\pi - \pi)}{\lambda - 1} - \frac{\sin(\lambda\pi + \pi)}{\lambda + 1} \right]$$

$$= -\frac{\sin(\lambda\pi)}{\pi} \left[\frac{1}{\lambda - 1} - \frac{1}{\lambda + 1} \right] = -\frac{2 \sin(\lambda\pi)}{\pi(\lambda^2 - 1)}$$

Note: $\sin(\lambda\pi \pm \pi) = -\sin(\lambda\pi)$.

25. Integration by parts gives the integral formula found in the Mathematical References, item 4.14. At the upper limit, the numerator of that formula approaches 0.

27. Integrate with respect to x, and reverse the order of integration to find

$$\frac{2}{\pi} \int_0^\infty \int_0^t e^{-\lambda} \cos(\lambda x)dxd\lambda = \int_0^t \frac{1}{1 + x^2}dx$$

$$\frac{2}{\pi} \int_0^\infty e^{-\lambda} \left(\frac{\sin(\lambda x)}{\lambda} \Bigg|_0^t \right) d\lambda = \tan^{-1}(x)\Bigg|_0^t$$

29. See the Answers.

31. There are infinitely many correct solutions. The easiest come from using period 2 for the series and assuming value 0 if $x > 1$ for the integrals. For efficiency, compute

$$S(\lambda) = \int_0^1 (1 - x) \sin(\lambda x) dx = \frac{1}{\lambda} - \frac{\sin(\lambda)}{\lambda^2}$$

$$C(\lambda) = \int_0^1 (1 - x) \cos(\lambda x) dx = \frac{1 - \cos(\lambda)}{\lambda^2}$$

Then the sine and cosine series coefficients are

$$b_n = 2S(n\pi) = \frac{2}{n\pi}; \quad a_n = 2C(n\pi) = \frac{2(1 - \cos(n\pi))}{(n\pi)^2}.$$

By geometry, $a_0 = \frac{1}{2}$. The sine and cosine integral coefficient functions are

$$B(\lambda) = \frac{2}{\pi} S(\lambda), \quad A(\lambda) = \frac{2}{\pi} C(\lambda).$$

33. See Excel worksheet.

35. Use integration by parts. After evaluation, terms containing $\sin(2n\pi/3)$ cancel out.

x	0	$a/3$	a	$-a/2$
sum	0	0	$a/3$	$a/6$

37. Use integration by parts. After evaluation, the terms containing $\sin(n\pi/2)$ cancel out.

x	0	$a/2$	a	$3a/2$
sum	0	1	$1/2$	1

39. Use integration by parts.

x	0	a	$-a/2$
sum	1	0	$1/2$

41. The easiest way to do the integration is the second integration-by-parts formula:

$$\int uv'' dx = uv' - u'v + \int vu'' dx$$

with $u = x(a - x)$ and $v = -\dfrac{\cos(n\pi x/a)}{(n\pi/a)^2}$

x	0	$-a$	$-a/2$
sum	0	0	$a^2/4$

43.

x	$-a$	$a/2$	a
sum	1	1/2	1

45.

x	$2a$	0	$-a$
sum	0	0	0

47. After evaluation the terms containing $\cos(n\pi/2)$ cancel.

x	0	0	$-a/2$
sum	0	0	$-a/2$

49.

x	0	$a/4$	$a/2$	a	$-3a/4$
sum	0	1/2	1	0	$-1/2$

51. Use the Integral 4.13 in the Mathematical References.

x	0	$a/2$	a	$-a$
sum	0	$e^{ka/2}$	0	0

53. Use the Integral 4.14 in the Mathematical References. At the upper limit, the evaluation gives 0.

55. See the Answers.

57. See the Answers.

59. Use the Integral 4.13 in the Mathematical References. At the upper limit the evaluation gives 0.

61. See the Answers.

63. See the Answers.

65. Abbreviate the nth term: $T_n = a_n \cos(nx) + b_n \sin(nx)$. Then

$$S_1 + \cdots + S_N = Na_0 + T_1 + (T_1 + T_2) + (T_1 + T_2 + T_3) + \cdots + (T_1 + T_2 + \cdots + T_N).$$

In this sum, T_1 appears N times, T_2 appears $N-1$ times,... T_n appears $N+1-n$ times. Thus,

$$\sigma_N = \frac{1}{N}(S_1 + \cdots + S_N) = a_0 + T_1 + \frac{N-1}{N}T_2 + \cdots + \frac{N+1-n}{N}T_n + \cdots + \frac{1}{N}T_N.$$

67. From Equation (13) of Section 1.7,

$$S_n(x) - f(x) = \frac{1}{\pi} \int_{-\pi}^{\pi} (f(x+y) - f(x)) \frac{\sin(n + \frac{1}{2})y}{2\sin(n/2)} \, dn.$$

Now add up for $n = 1$ to N and divide by N. Since $f(x)$ does not depend on n, it remains unchanged:

$$\sigma_N(x) - f(x) = \frac{1}{N\pi} \int_{-\pi}^{\pi} (f(x+y) - f(x)) \sum_{n=1}^{N} \frac{\sin((n + \frac{1}{2})y)}{2\sin(y/2)} \, dn$$

$$\frac{1}{N\pi} \int_{-\pi}^{\pi} (f(x+y) - f(x)) \frac{1}{2\sin(y/2)} \cdot \frac{\sin^2(Ny/2)}{\sin(y/2)} \, dn$$

(using the identity in Exercise 66).

69. a. Set $x = 0$, so that

$$0 = \frac{1}{2} - \frac{4}{\pi^2} \sum_{k=0}^{\infty} \frac{1}{(2k+1)^2}$$

Isolate the series algebraically

$$\frac{\pi^2}{8} = \sum_{k=0}^{\infty} \frac{1}{(2k+1)^2} = 1 + \frac{1}{9} + \frac{1}{25} + \cdots$$

b. Set $x = 1/2$, giving

$$\frac{4}{\pi} \sum_{k=0}^{\infty} \frac{1}{2k+1} \sin((2k+1)\pi/2) = 1.$$

Now, $\sin((2k+1)\pi/2) = 1$ (k even) or $= -1$ (k odd). Transfer $4/\pi$ to the other side and write out some terms of the series:

$$1 - \frac{1}{3} + \frac{1}{5} - + \cdots = \frac{\pi}{4}$$

c. Set $x = 0$; isolate the series algebraically.

Chapter 2

In some places, we will be using subscript notation for partial derivatives.

$$u_x \quad \text{means} \quad \frac{\partial u}{\partial x}$$

$$u_{xx} \quad \text{means} \quad \frac{\partial^2 u}{\partial x^2}$$

$$u_t \quad \text{means} \quad \frac{\partial u}{\partial t}$$

In boundary and initial conditions we have occasion to write symbols like $\frac{\partial u}{\partial x}(0, t)$ or $u_x(0, t)$. In both of these, the indicated differentiation is done first, then the evaluation at $x = 0$.

2.1 Derivation and Boundary Conditions

See Answers.

Chapter 2

2.2 Steady-State Temperatures

1. The steady-state problem is

$$\frac{d^2v}{dx^2} - \gamma^2(v - U) = 0, \quad 0 < x < a$$

$$v(0) = T_0, \quad v(a) = T.$$

In standard form, the differential equation is $v'' - \gamma^2 v = -\gamma^2 U$. This is linear and nonhomogeneous. A particular solution is constant, $v_p = A$, and $A = U$ in order to satisfy the differential equation. The homogeneous equation $v'' - \gamma^2 v = 0$ has a general solution in terms of $e^{\gamma x}$ and $e^{-\gamma x}$ or $\cosh(\gamma x)$ and $\sinh(\gamma x)$. The latter pair are more convenient in the finite interval.

The general solution of the differential equation is $v(x) = U + c_1 \cosh(\gamma x) + c_2 \sinh(\gamma x)$. Now, apply the boundary conditions:

$$u(0) = T_0 : \quad U + c_1 = T_0$$

$$v(a) = T_1 : \quad U + c_1 \cosh(\gamma a) + c_2 \sinh(\gamma a) = T_1.$$

Solve these for the coefficients

$$c_1 = T_0 - U, \quad c_2 = \frac{T_1 - U - (T_0 - U)\cosh(\gamma a)}{\sinh(\gamma a)}.$$

3. The steady-state problem is

$$v'' + \gamma^2(u - T) = 0, \quad 0 < x < a,$$

$$v(0) = T, \quad v(a) = T.$$

The solution process is similar to Exercise 1 shown above, except that the homogeneous equation is $v'' + \gamma^2 v = 0$, with $\cos(\gamma x)$ and $\sin(\gamma x)$ as independent solutions. The general solution of the differential equation is

$$v(x) = T + c_1 \cos(\gamma x) + c_2 \sin(\gamma x).$$

The boundary conditions are

$$v(0) = T : \quad T + c_1 = T$$

$$v(a) = T : \quad T + c_1 \cos(\gamma a) + c_2 \sin(\gamma a) = T$$

Hence, $c_1 = c_2 = 0$, and $v(x) = T$. However if $\gamma = \pi/a$, $\sin(\gamma a) = \sin(\pi) = 0$, so c_2 is not determined by the boundary conditions.

5. The steady-state problem is

$$\frac{d}{dx}\left((\kappa_0 + \beta x)\frac{dv}{dx}\right) = 0, \quad 0 < x < a$$

$$v(0) = T_0, \quad v(a) = T_1.$$

To solve the differential equation, see that $(\kappa_0 + \beta x)v' = c_1$, $v' = c_1/(\kappa_0 + \beta x)$, $v = (c_1/\beta)\ln(\kappa_0 + \beta x) + c_2$. Now apply boundary conditions.

$$v(0) = T_0: \quad \frac{c_1}{\beta}\ln(\kappa_0) + c_2 = T_0$$

$$v(a) = T_1: \quad \frac{c_1}{\beta}\ln(\kappa_0 + \beta a) + c_2 = T_1$$

Subtract to find

$$\frac{c_1}{\beta}(\ln(\kappa_0 + \beta a) - \ln(\kappa_0)) = T_1 - T_0$$

or

$$\frac{c_1}{\beta} = \frac{(T_1 - T_0)}{\ln\left(1 + \dfrac{\beta a}{\kappa_0}\right)},$$

and $c_2 = T_0 - \left(\dfrac{c_1}{\beta}\right)\ln \kappa_0$.

7. The steady-state problem is

$$v'' + r = 0, \quad 0 < x < a$$

$$v(0) = T_0, \quad v'(a) = 0$$

The general solution of the differential equation (found easily by integration, since r is constant) is $v(x) = c_1 + c_2 x - rx^2/2$. Apply the boundary conditions

$$v(0) = T_0: \quad c_1 = T_0$$

$$v'(a) = 0: \quad c_2 - ra = 0$$

Hence, $v(x) = T_0 + rax - rx^2/2$.

9. The steady-state problem is

$$Dv'' = Sv', \quad 0 < x < a$$

$$v(0) = U, \quad v(a) = 0$$

In standard form, the differential equation is $v'' - \gamma v' = 0$ (using $S/D = \gamma$) and its solution is $v(x) = c_1 + c_2 e^{\gamma x}$. Apply the boundary conditions

$$v(0) = U: \quad c_1 + c_2 = U$$

$$v(a) = 0: \quad c_1 + c_2 e^{\gamma a} = 0$$

Subtract to find $c_2(e^{\gamma a} - 1) = -U$, $c_1 = -c_2 e^{\gamma a} = Ue^{\gamma a}/(e^{\gamma a} - 1)$. Hence $v(x) = U(e^{\gamma a} - e^{\gamma x})/(e^{\gamma a} - 1)$.

Chapter 2

2.3 Example: Fixed End Temperatures

1. See the Answers.

3.

$$\frac{\partial u}{\partial t} = \frac{\partial U}{\partial \tau} \cdot \frac{d\tau}{dt} = \frac{\partial U}{\partial \tau} \cdot \frac{k}{a^2},$$

$$\frac{\partial u}{\partial x} = \frac{\partial U}{\partial \xi} \cdot \frac{d\xi}{dx} = \frac{\partial U}{\partial \xi} \cdot \frac{1}{a}, \quad \frac{\partial^2 u}{\partial x^2} = \frac{\partial^2 U}{\partial \xi^2} \cdot \frac{1}{a^2}.$$

The partial differential equation (8) becomes

$$\frac{\partial^2 U}{\partial \xi^2} \frac{1}{a^2} = \frac{1}{k} \frac{\partial U}{\partial \tau} \cdot \frac{k}{a^2}$$

or

$$\frac{\partial^2 U}{\partial \xi^2} = \frac{\partial U}{\partial \tau}, \quad 0 < \xi < 1, \quad 0 < \tau.$$

5. Equation (15) gives the solution of the heat equation that satisfies the boundary conditions. The initial condition becomes

$$\sum_{n=1}^{\infty} b_n \sin\left(\frac{n\pi x}{a}\right) = T_0, \quad 0 < x < a,$$

so

$$b_n = \frac{2}{a} \int_0^a T_0 \sin\left(\frac{n\pi x}{a}\right) dx = \frac{2T_0(1 - \cos(n\pi))}{n\pi}$$

7. Use Equation (15). The initial condition becomes

$$\sum_{n=1}^{\infty} b_n \sin\left(\frac{n\pi x}{a}\right) = \beta(a - x), \quad 0 < x < a,$$

so

$$b_n = \frac{2}{a} \int_0^a \beta(a - x) \sin\left(\frac{n\pi x}{a}\right) dx$$

$$= \frac{2\beta}{a} \left[-\frac{(a - x)\cos\left(\frac{n\pi x}{a}\right)}{\frac{n\pi}{a}} \Bigg|_0^a + \frac{\sin\left(\frac{n\pi x}{a}\right)}{\left(\frac{n\pi}{a}\right)^2} \Bigg|_0^a \right] = \frac{2\beta}{n\pi}.$$

9.a The steady-state problem is

$$v'' = 0, \quad v(0) = C_1, \quad v(a) = C_1,$$

with solution $v(x) = C_1$.

b. See Answers.

c. Use Equation (15) and (replacing T_0 with $C_0 - C_1$) the solution of Exercise 5 above to find

$$b_n = \frac{2(C_0 - C_1)(1 - \cos(n\pi))}{n\pi}$$

d. Since $C(x, t) = C_1 + w(x, t)$, $C(x, t) - C_0 = C_1 - C_0 + w(x_1 t)$. When the left side is $0.9(C_1 - C_0)$ at $x = a/2$, we have

$$0.9(C_1 - C_0) = C_1 - C_0 + w\left(\frac{a}{2}, t\right).$$

Now, set $x = a/2$ in the first term of the series for w

$$-0.1(C_1 - C_0) = 2(C_0 - C_1) \cdot \frac{2}{\pi} e^{-(\pi/a)^2 Dt}$$

$$\frac{\pi}{4}(0.1) = e^{-(\pi/a)^2 D}t, (\pi/a)^2 Dt = -\ln(\pi/40) = 2.544$$

e.

$$t = \frac{2.544a^2}{\pi^2 D} = \frac{2.544 \cdot (5 \times 10^{-6})^2 m^2}{\pi^2 \cdot 10^{-11} cm^2/s}$$

$$= 6.444 \times 10^{-1} m^2/cm^2$$

But $1m = 100cm$, so $(m/cm)^2 = 10^4$ and $t = 6.444 \times 10^3$ s.

Chapter 2

2.4 Example: Insulated Bar

1. Equation (9) gives the solution of the problem, and Equation (10) identifies the coefficients. If $f(x) = T_1 x/a$, then $a_0 = T_1/2$ and

$$a_n = \frac{2}{2} \int_0^a \frac{T_1 x}{a} \cos\left(\frac{n\pi x}{a}\right) dx$$

$$= \frac{2T_1}{a^2}\left[\frac{\cos\left(\frac{n\pi x}{a}\right)}{\left(\frac{n\pi}{a}\right)^2} + \frac{x\sin\left(\frac{n\pi x}{a}\right)}{\frac{n\pi}{a}}\right]\Bigg|_0^a = \frac{2T_1}{\pi^2}\frac{\cos(n\pi) - 1}{n^2}.$$

Since $a_2 = 0$, the approximate solution required is

$$w(x, t) \cong \frac{T_1}{2} - \frac{2T_1}{\pi^2}\left[2\cos(\frac{\pi x}{a})e^{-(\pi/a)^2 kt} + \frac{2}{9}\cos\left(\frac{3\pi x}{a}\right)e^{-9(\pi/a)^2 kt}\right]$$

The figure shows this function (for $T_1 = 100$, $a = 1$) at times such that $kt/a^2 = 0.01, 0.1$ and 0.5.

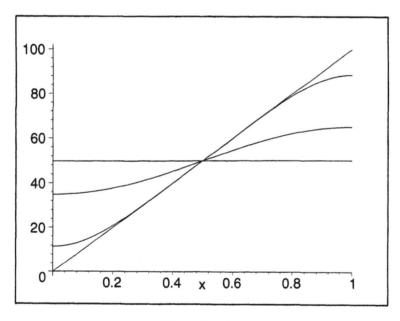

Exercise 1

3. The solution is given in Equations (9) and (10). The function is a triangle with maximum T_0 at $x = a/2$. The coefficients are $a_0 = T_0/2$ and $a_n = 4T_0\dfrac{2\cos(n\pi/2) - \cos(n\pi) - 1}{n^2\pi^2}$.

5.a. The steady-state problem is

$$v'' = 0, \qquad v'(0) = S_0, \qquad v'(a) = S_1.$$

The general solution of the differential equation is $v(x) = c_1 + c_2 x$. Apply the boundary conditions.

$$v'(0) = S_0: \quad c_2 = S_0$$

$$v'(a) = S_1 : \quad c_2 = S_1$$

These are inconsistent unless $S_0 = S_1$. If this is the case, c_1 remains undetermined. If $S_0 \neq S_1$, there is a net flow of heat into or out of the object.

b. If $S_0 = S_1$, then set $u(x,t) = S_0 x + w(x,t)$. The function $w(x,t)$ satisfies the heat equation because

$$\frac{\partial^2 u}{\partial x^2} = \frac{\partial^2 w}{\partial x^2}, \quad \frac{\partial u}{\partial t} = \frac{\partial w}{\partial t}.$$

The boundary conditions are

$$S_0 + \frac{\partial w}{\partial x}(0,t) = S_0 \quad \text{or} \quad \frac{\partial w}{\partial x}(0,t) = 0$$

and similarly at $x = a$.

c. For $u = A(kt + x^2/2) + Bx$, compute $\frac{\partial u}{\partial t} = Ak$ and $\frac{\partial^2 u}{\partial x^2} = A$. Thus u satisfies the heat equation. Since $\frac{\partial u}{\partial x} = Ax + B$, the boundary conditions become

$$\frac{\partial u}{\partial x}(0,t) = S_0 : \quad B = S_0$$

$$\frac{\partial u}{\partial x}(a,t) = S_1 : \quad Aa + B = S_1$$

If $S_1 \neq S_2$, then $A \neq 0$ and the given function contains the term At.

7. The eigenvalue problem is

$$\phi'' + \lambda^2 \phi = 0, \quad 0 < x < a,$$

$$\phi'(0) = 0, \quad \phi'(a) = 0.$$

The solution is $\lambda_0 = 0$, $\lambda_n = n\pi/a$, $(n = 1, 2, \cdots)$ with corresponding eigenfunctions

$$\phi_0(x) = 1, \quad \phi_n(x) = \cos(\lambda_n x).$$

9. $A_n(t_1) = a_n \exp\left(-\left(\frac{n\pi}{a}\right)^2 kt_1\right) = a_n r^{-n^2}$ with $r = \exp\left(-(\frac{\pi}{a})^2 kt_1\right)$. Then since $a_n \to 0$, there is a constant A for which $|a_n| \leq A$ for all n. Finally $r < 1$ and $|A_n(t_1)| \leq Ar^{-n^2} \leq Ar^{-n}$. Therefore $\sum_1^\infty |A_n(t_1)|$ converges by comparison to a geometric series. The series for $\frac{\partial^2 u}{\partial x^2}$ has coefficients $-(\frac{n\pi}{a})^2 A_n(t_1)$ and the series $\sum |(\frac{n\pi}{a})^2 A_n(t_1)|$ converges by comparison with the second derivative of the geometric series.

11. No, $u(0,t)$ is not constant. The condition that would imply that $u(0,t)$ is constant is: $\frac{\partial u}{\partial t}(0,t) = 0$, $0 < t$. This is quite different from the zero-slope condition $\frac{\partial u}{\partial x}(0,t) = 0$, $0 < t$.

13. Check the partial differential equation by differentiating the series term by term.

Chapter 2

2.5 Example: Different Boundary Conditions

1. The steady-state problem is

$$v'' = 0, \quad v(0) = T_0, \quad v'(a) = 0$$

The general solution of the differential equation is $v(x) = c_1 + c_2 x$. The boundary conditions give

$$v(0) = T_0: \quad c_1 = T_0$$
$$v'(a) = 0: \quad c_2 = 0$$

3. The steady-state solution is $v(x) = T_0$. The initial condition for the transient is $w(x, 0) = Tx/a - T_0$. Then Equation (16) gives the solution of the transient problem with coefficients from Equation (18):

$$b_n = \frac{2}{a} \int_0^a \left(\frac{Tx}{a} - T_0 \right) \sin(\lambda_n x) dx$$

$$= \frac{2}{a} \left[\frac{T}{a} \left(\frac{\sin(\lambda_n x)}{\lambda_n^2} - \frac{x \cos(\lambda_n x)}{\lambda_n} \right) \Big|_0^a + T_0 \frac{\cos \lambda_n x}{\lambda_n} \Big|_0^a \right] = \frac{2}{a} \left[\frac{T}{a} \frac{\sin(\lambda_n a)}{\lambda_n^2} - \frac{T_0}{\lambda_n} \right]$$

In the evaluation, recall that $\cos(\lambda_n a) = 0$ and $\sin(\lambda_n a) = (-1)^{n+1}$.

5. The steady-state problem is $v'' = -T/a^2$, $v(0) = T_0$, $v'(a) = 0$. The general solution of the differential equation is $v(x) = c_1 + c_2 x - \frac{Tx^2}{2a^2}$. Boundary conditions:

$$v(0) = T_0: \quad c_1 = T_0$$

$$v'(a) = 0: \quad c_2 - T/a = 0$$

Hence $v(x) = T_0 + T \left(\frac{x}{a} - \frac{x^2}{2a^2} \right)$. The transient problem is the same as Equations (5)-(8) with $g(x) = \frac{Tx(x - 2a)}{2a^2}$. Then by Equation (18)

$$b_n = \frac{2}{a} \int_0^a \frac{Tx(x - 2a)}{2a^2} \sin(\lambda_n x) dx = \frac{T}{a^3} \int_0^a (x^2 - 2ax) \sin(\lambda_n x) dx \qquad *$$

$$= \frac{T}{a^3} \left[-\frac{\cos(\lambda_n x)}{\lambda_n} (x^2 - 2ax) \Big|_0^a + \frac{\sin \lambda_n x}{\lambda_n^2} (2x - 2a) \Big|_0^a + \int_0^a -\frac{\sin \lambda_n x}{\lambda_n^2} \cdot 2 dx \right]$$

$$= \frac{T}{a^3} \left[2 \frac{\cos \lambda_n x}{\lambda_n^3} \Big|_0^a \right] = -\frac{2T}{(\lambda_n a)^3}$$

* Use the second integration-by-parts formula, Mathematical References, Calculus 4b, with $u = x^2 - 2ax$, $v = -\frac{\sin(\lambda_n x)}{\lambda_n^2}$. Recall that $\cos(\lambda_n a) = 0$.

7. Steady state problem: $v'' = 0$, $v'(0) = 0$, $v(a) = T_0$. Solution: $v(x) = T_0$. Transient problem:

$$\frac{\partial^2 w}{\partial x^2} = \frac{1}{k}\frac{\partial w}{\partial t}, \quad 0 < x < a, \quad 0 < t$$

$$\frac{\partial w}{\partial x}(0, t) = 0, \quad w(a, t) = 0,$$

$$w(x, 0) = T_1 - T_0, \quad 0 < x < a$$

Eigenvalue problem $\phi'' + \lambda^2 \phi = 0$, $\phi'(0) = 0$, $\phi(a) = 0$. Solutions $\phi_n(x) = \cos(\lambda_n x)$, $\lambda_n = \dfrac{(n - \frac{1}{2})\pi}{a}$, $n = 1, 2, \cdots$. Solution of transient problem

$$w(x, t) = \sum_{n=1}^{\infty} c_n \cos(\lambda_n x) \exp\left(-\lambda_n^2 k t\right)$$

$$c_n = \frac{2}{a}\int_0^a (T_1 - T_0)\cos(\lambda_n x)dx = \frac{2}{a}(T_1 - T_0)\frac{\sin(\lambda_n a)}{\lambda_n}$$

9. Since $u(a, t) = 0$, the problem is homogeneous, so the formula for $w(x, t)$ in Exercise 7 is the formula for $u(x, t)$ in this one. The function in the initial condition is one of the eigenfunctions, so

$$u(x, t) = T_1 \cos\left(\frac{\pi x}{2a}\right)\exp\left(-\left(\frac{\pi}{2a}\right)^2 kt\right).$$

That is, $c_1 = T$, and the rest of the c's are 0.

11. See Answers

13. See Answers.

15. See Answers.

Chapter 2

2.6 Example: Convection

1. Note that $v(a)$ is always between T_0 and T_1.

3. Negative values of λ produce no new eigenfunctions, because λ^2, and not λ, appears in the differential equation.

5. See Figure.

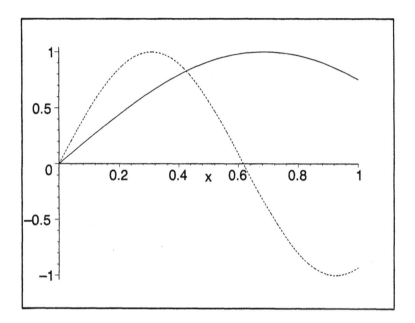

7. Use Equation (14)

$$\int_0^a g(x)\sin(\lambda_m x)dx = \int_0^a \sin(\lambda_m x)dx = \frac{1 - \cos(\lambda_m a)}{\lambda_m}$$

$$\int_0^a \sin^2(\lambda_m x)dx = \frac{x}{2} - \left.\frac{\sin(2\lambda_m x)}{4\lambda_m}\right|_0^a = \frac{a}{2} - \frac{\sin(2\lambda_m a)}{4\lambda_m}$$

Use the relation in Equation (11) to replace $\cos(\lambda_m a) = -\dfrac{h\sin(\lambda_m a)}{\kappa\lambda_m}$.

$$b_m = 2\frac{1 + \dfrac{h\sin(\lambda_m a)}{\kappa\lambda_m}}{\lambda_m\left(a - \dfrac{h\sin^2(\lambda_m a)}{\kappa\lambda_m^2}\right)}$$

(also using $\sin(2A) = 2\sin(A)\cos(A)$). This form makes it clear that $b_m \cong 2/(\lambda_m a)$ for large m.

9.

$$\int_0^a x\sin(\lambda_m x)dx = \left[\frac{\sin(\lambda_m x)}{\lambda_m^2} - \frac{x\cos(\lambda_m x)}{\lambda_m}\right]_0^a = \frac{\sin(\lambda_m a)}{\lambda_m^2} - \frac{a\cos(\lambda_m a)}{\lambda_m}$$

$$= \frac{\sin(\lambda_m a)}{\lambda_m^2} \left(1 + \frac{ah}{\kappa} \right).$$

See the solution of #7.

$$b_m = 2 \frac{\sin(\lambda_m a) \left(1 + \dfrac{ah}{\kappa} \right)}{\lambda_m^2 \left(a - \dfrac{h \sin^2(\lambda_m a)}{\kappa \lambda_m^2} \right)}$$

Chapter 2

2.7 Sturm-Liouville Problems

1. Apply the boundary conditions:

$$\phi(1) = 0: \quad c_1 = 0$$

$$\phi(2) = 0: \quad c_2 \sin(\lambda \ln(2)) = 0$$

$\lambda_n = n\pi/\ln(2)$, and $\phi_n = \sin(\lambda_n \ln(x))$. The orthogonality relation is

$$\int_1^2 \sin\left(\lambda_n \ln(x)\right) \sin\left(\lambda_m \ln(x)\right) \cdot \frac{1}{x} dx = 0.$$

To prove it, change variable to $y = \ln(x)$. The integral becomes

$$\int_0^{\ln(2)} \sin\left(\frac{n\pi y}{\ln(2)}\right) \sin\left(\frac{m\pi y}{\ln(2)}\right) dy$$

which is 0 by direct integration if $n \neq m$.

3. The general solution of the differential equation is

$$\phi(x) = c_1 \cos(\lambda x) + c_2 \sin(\lambda x).$$

Apply the boundary conditions to find a condition on λ that allows a non-zero solution. For each case, the equations are shown that come from the boundary conditions.

a. $c_1 = 0$, $\quad c_2 \lambda \cos(\lambda a) = 0$. $\quad \lambda_n = (n - \frac{1}{2})\pi/a$

b. $c_2 = 0$, $\quad c_1 \lambda \cos(\lambda a) = 0$.

c. $c_1 = 0$, $\quad c_2(\sin(\lambda a) + \lambda \cos(\lambda a)) = 0$. λ_n is the nth solution of $\sin(\lambda a) + \lambda \cos(\lambda a) = 0$.

d. $c_1 - \lambda c_2 = 0$, $\lambda(-c_1 \sin(\lambda a) + c_2 \cos(\lambda a)) = 0$. Use $c_1 = \lambda c_2$. The second equation becomes $c_2(-\lambda \sin(\lambda a) + \cos(\lambda a)) = 0$. λ_n is the nth solution (0 is not an eigenvalue), and

$$\phi_n(x) = \lambda_n \cos(\lambda_n x) + \sin(\lambda_n x)$$

(or any multiple).

e. $c_1 - \lambda c_2 = 0$, $c_1(\cos(\lambda a) - \lambda \sin(\lambda a)) + c_2(\sin(\lambda a) + \lambda \cos(\lambda a)) = 0$. Use $c_1 = \lambda c_2$ from the first equation and sort terms to find $(1 - \lambda^2) \sin(\lambda a) + 2\lambda \cos(\lambda a) = 0$. Then λ_n is the nth solution of this equation, and $\phi_n(x) = \lambda_n \cos(\lambda_n x) + \sin(\lambda_n x)$ (or any multiple).

5. In each case, the differential equation is in the form required by Equation (5). You can just read off the weight function.

a. $p(x) = 1 + x$

b. $p(x) = e^x$

c. $p(x) = 1/x^2$. Note that the differential equation is not singular because $x = 0$ is not in the interval.

d. $p(x) = e^x$. Note that $\sin(x) = q(x)$ for Equation (5).

7. The appearance of λ^2 in the boundary condition means that condition d. in the definition of regular Sturm-Liouville problem is not met. The eigenfunctions can be found without difficulty as $\phi_n(x) = \sin(\lambda_n x)$, with λ_n a solution of $\lambda \cos(\lambda a) - \lambda^2 \sin(\lambda a) = 0$ ($\lambda = 0$ is not an eigenvalue.)

To show non-orthogonality:

$$\int_0^a \sin(\lambda_n x) \sin(\lambda_m x) dx = \frac{\sin((\lambda_n - \lambda_m)a)}{2(\lambda_n - \lambda_m)} - \frac{\sin((\lambda_n + \lambda_m)a)}{2(\lambda_n + \lambda_m)}$$

First, use the identities

$$\sin(\lambda_n a \pm \lambda_m a) = \sin(\lambda_n a)\cos(\lambda_m a) \pm \sin(\lambda_m a)\cos(\lambda_n a)$$

Next, use the eigenvalue equation: $\cos(\lambda a) = \lambda \sin(\lambda a)$ to eliminate cosines

$$\sin(\lambda_n a \pm \lambda_m a) = \lambda_m \sin(\lambda_n a)\sin(\lambda_m a) \pm \lambda_n \sin(\lambda_m a)\sin(\lambda_n a)$$

$$= (\lambda_m \pm \lambda_n)\sin(\lambda_n a)\sin(\lambda_m a).$$

Now the value of the integral is

$$\left[\frac{\lambda_m - \lambda_n}{2(\lambda_n - \lambda_m)} - \frac{\lambda_m + \lambda_n}{2(\lambda_n + \lambda_m)}\right]\sin(\lambda_n a)\sin(\lambda_m a) = -\sin(\lambda_n a)\sin(\lambda_m a).$$

Neither of the factors is 0.

9. The solution of the problem is routine if $\mu = \lambda^2$. If $\mu = -p^2$, the general solution of the differential equation is

$$\phi(x) = c_1 \cosh(px) + c_2 \sinh(px)$$

and the boundary conditions require: $c_1 + pc_2 = 0$,

$$c_1 \left(\cosh(pa) + p\sinh(pa)\right) + c_2 \left(\sinh pa + p\cosh(pa)\right) = 0.$$

Use $c_1 = -pc_2$ and find that $(1 - p^2)\sinh(pa) = 0$. Therefore $p^2 = 1$ and $\mu = -1$.

Theorem 2 requires the left boundary condition of the form $\alpha_1 \phi(0) - \alpha_2 \phi'(0) = 0$ with $\alpha_1 \geq 0$, $\alpha_2 \geq 0$, but we have $\alpha_1 = 1$, $\alpha_2 = -1$. The Theorem is not contradicted.

Chapter 2

2.8 Expansion in Series of Eigenfunctions

1. See Example 2 and Exercise 1 of Section 2.7. From the convergence theorem, the numerator of the ratio for c_n is

$$\int_1^b x \sin(\lambda_n \ln(x)) \frac{dx}{x} = \int_0^\beta e^y \sin(\lambda_n y) dy$$

with $y = \ln(x)$, $\beta = \ln(b)$. This integral is

$$\frac{e^y(\sin(\lambda_n y) - \lambda_n \cos(\lambda_n y))}{\lambda_n^2 + 1} \bigg|_0^\beta = \frac{-e^\beta \lambda_n \cos(\lambda_n \beta) + \lambda_n}{\lambda_n^2 + 1} = \frac{\lambda_n(1 - b\cos(n\pi))}{\lambda_n^2 + 1}$$

Use the fact that $\lambda_n \beta = n\pi$ and $e^\beta = b$. The denominator of the ratio for c_n is $\beta/2$. Thus

$$c_n = \frac{2}{\ln(b)} \frac{\lambda_n(1 - b\cos(n\pi))}{\lambda_n^2 + 1}.$$

See the Figure for the sum of 20 terms of the series.

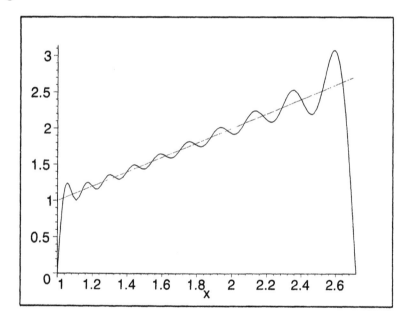

3. Differentiate the product in the differential equation. Then every term has an exponential, which factors out, leaving

$$\phi'' + \phi' + \gamma^2 \phi = 0$$

The general solution of the differential equation is

$$\phi(x) = e^{-x/2}(c_1 \cos(\lambda x) + c_2 \sin(\lambda x))$$

with $\lambda = \sqrt{\gamma^2 - 1}$. The boundary conditions require $\lambda_n = n\pi/a$.

The weight function is $p(x) = e^x$, so

$$c_n = \frac{2}{a} \int_0^a f(x) e^{x/2} \sin(\lambda_n x).$$

For $f(x) = 1$, $0 < x < a$,

$$c_n = \frac{2}{a} \int_0^a e^{x/2} \sin\left(\frac{n\pi x}{a}\right) dx.$$

(See the Answers.) See Figure for the sum of 20 terms of the series.

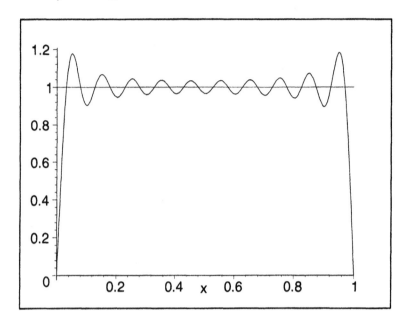

5. See Answers. The denominator of the formula for c_n in the Theorem is just 1 because the functions are normalized.

7. $\phi_0(x) = 1$ and $\phi_n(x) = \cos(n\pi x)$ for $n = 1, 2, \cdots$. Since

$$\int_0^1 \phi_0^2(x)dx = 1 \quad \text{and} \quad \int_0^1 \cos^2(n\pi x)dx = \frac{1}{2},$$

The normalized eigenfunctions are $\Psi_0(x) = 1$, $\quad \Psi_n(x) = \sqrt{2}\cos(n\pi x)$.

Chapter 2

2.9 Generalities on the Heat Conduction Problem

See Answers.

2.10 Semi-Infinite Rod

1. See Answers

3. See Answers

5. The analogs of Equation (9) and (10) for this case are

$$u(x,t) = \int_0^\infty A(\lambda)\cos(\lambda x)e^{-\lambda^2 kt}d\lambda \tag{9'}$$

$$A(\lambda) = \frac{2}{\pi}\int_0^\infty f(x)\cos(\lambda x)dx \tag{10'}$$

If $f(x)$ is as in Exercise 1,

$$A(\lambda) = \frac{2}{\pi}\int_a^b T\cos(\lambda x)dx = \frac{2T}{\pi}\sin(\lambda) = \frac{2T}{\pi}\frac{(\sin(\lambda b) - \sin(\lambda a))}{\lambda}.$$

7. This problem has a steady-state solution $\lim_{t\to\infty} u(x,t) = T_0$. Then $w(x,t) = u(x,t) - T_0$ satisfies the problem in Equation (1) and (2), and Equation (3) is replaced by $w(x,0) = f(x) - T_0$. The solution given in Equation (9) requires that

$$B(\lambda) = \frac{2}{\pi}\int_0^\infty (f(x) - T_0)\sin(\lambda x)dx. \tag{9''}$$

We need $\int_0^\infty |f(x) - T_0|dx \to \infty$. If this is the case, then

$$u(x,t) = T_0 + \int_0^\infty B(\lambda)\sin(\lambda x)e^{-\lambda^2 kt}d\lambda$$

with $B(\lambda)$ as in Equation (9'').

9.a. The steady-state problem is
$$v'' - a^2 v = 0, \quad 0 < x$$
$$v(0) = C_0.$$

The solution is $v(x) = C_0 e^{-ax}$.

b. The problem for the transient is

$$\frac{\partial^2 w}{\partial x^2} = \frac{1}{D}\frac{\partial w}{\partial t} + a^2 w, \quad 0 < x$$

$$w(0,t) = 0, \quad w(x,0) = -C_0 e^{-ax}$$

c. If $w(x,t) = \phi(x)T(t)$, then
$$\frac{\phi''(x)}{\phi(x)} = \frac{T'(t)}{DT(t)} + a^2 = -\lambda^2.$$

With the boundary condition $\phi(0) = 0$ (and the boundedness condition) we have $\phi(x;\lambda) = \sin(\lambda x)$ and $T(t) = e^{-(a^2+\lambda^2)Dt}$.

The solution formed by combining the product solutions is

$$w(x,t) = \int_0^\infty B(\lambda)e^{-(a^2+\lambda^2)Dt}\sin(\lambda x)d\lambda.$$

The initial condition requires

$$-C_0e^{-ax} = \int_0^\infty B(\lambda)\sin(\lambda x)d\lambda.$$

Hence, from Fourier Integral theory

$$B(\lambda) = \frac{2}{\pi}\int_0^\infty -C_0e^{-ax}\sin(\lambda x)dx = -\frac{2}{\pi}C_0\frac{\lambda}{a^2+\lambda^2}$$

Chapter 2

2.11 Infinite Rod

1. The integral in Equation (7) has to break at 0:

$$u(x,t) = \frac{1}{\sqrt{4k\pi t}} \left[\int_{-\infty}^{0} T_0 E(x') dx' + \int_{0}^{\infty} T_1 E(x') dx' \right]$$

where $E(x') = \exp\left[\frac{-(x'-x)^2}{4kt}\right]$ has been used for brevity.

3. The given function $f(x)$ is an even function, so that Equation (4) becomes

$$u(x,t) = \int_{0}^{\infty} A(\lambda) \cos(\lambda x) e^{-\lambda^2 kt} d\lambda$$

$$A(\lambda) = \frac{2}{\pi} \int_{0}^{\infty} T_0 e^{-x/a} \cos(\lambda x) dx = \frac{2}{\pi} T_0 \frac{1/a}{\lambda^2 + (1/a)^2}.$$

5. Differentiate and substitute into the differential equation:

$$\frac{\partial u}{\partial t} = \frac{1}{\sqrt{4k\pi}} \left[\frac{1}{\sqrt{t}} E \cdot \frac{x^2}{4kt^2} - \frac{\frac{1}{2}}{t^{\frac{3}{2}}} E \right] = \frac{E}{\sqrt{4k\pi}} \left[\frac{x^2}{4kt^{\frac{5}{2}}} - \frac{1}{2t^{\frac{3}{2}}} \right]$$

$$\left(\text{where} \quad E = \exp\left(\frac{-x^2}{4kt}\right) \right)$$

$$\frac{\partial u}{\partial x} = \frac{1}{\sqrt{4k\pi}} \cdot \frac{1}{\sqrt{t}} E \cdot \left(\frac{-2x}{4kt} \right)$$

$$\frac{\partial^2 u}{\partial x^2} = \frac{1}{\sqrt{4k\pi}} \cdot \frac{1}{\sqrt{t}} \left[E\left(\frac{-2}{4kt}\right) + E\left(\frac{-2x}{4kt}\right)^2 \right] = \frac{E}{\sqrt{4k\pi}} \left[\frac{-1}{2kt^{\frac{3}{2}}} + \frac{x^2}{4k^2 t^{\frac{5}{2}}} \right].$$

The relevant derivatives differ only by a factor of k.

7. Substitute directly into Equations (1)-(3) to see that $u(x,t) = 1$ is a solution. Now substitute $f(x') = 1$ into Equation (7). The solution must be the same either way; hence the equality.

9. To use Equation (4), the coefficient functions of Equation (5) have to be calculable. However, the integrals do not converge. On the other hand, choosing $A(\lambda) \equiv 0$ and $B(\lambda) = 2/(\pi\lambda)$ gives a solution.

Chapter 2

2.12 The Error Function

1. $\int_0^{-z} e^{-y^2} dy = -\int_0^{z} e^{-u^2} du$, by the substitution $u = -y$ and elementary properties of integrals.

3.a. Use Leibniz's rule (Mathematical References, Calculus, 3c).

b. $\text{erfc}(0) = 1 - \text{erf}(0) = 1$ (Equation (10))

c. $\lim_{z \to \infty} \text{erfc}(z) = 1 - \lim_{z \to \infty} \text{erf}(z) = 1 - 1.$

d. $\lim_{z \to -\infty} \text{erfc}(z) = 1 - \lim_{z \to -\infty} \text{erf}(z) = 1 - (-1).$

5. See Answers

7. Let $x = y^2$. Then

$$\int \frac{e^{-y^2}}{y} \cdot 2y\,dy = 2\int e^{-y^2} dy$$

Thus

$$I(x) = c + 2\int_0^{\sqrt{x}} e^{-y^2} dy = c + \sqrt{\pi}\,\text{erf}(\sqrt{x}).$$

9. $w(x,t) = u(x,t) - U_b$ satisfies the heat equation with conditions $w(0,t) = 0$, $w(x,0) = U_i - U_b$. Then $w(x,t) = (U_i - U_b)\text{erf}(\frac{x}{\sqrt{4kt}})$ by the developments of the text, and $u(x,t) = U_b + (U_i - U_b)\text{erf}(\frac{x}{\sqrt{4kt}})$.

Now set $u(x,t) = 0$ and solve for x as a function of t:

$$\text{erf}\left(\frac{x}{\sqrt{4kt}}\right) = -\frac{U_b}{U_i - U_b} > 0$$

Then $x = z\sqrt{4kt}$ if z is defined by $\text{erf}(z) = -\frac{U_b}{(U_i - U_b)}.$

Chapter 2

Miscellaneous

In Exercises 1-16 we give the steady-state problem (SSP), its solution, the transient problem (TP), the associated eigenvalue problem (EVP), its solution, the solution of the TP, and the values for any coefficients. Also see Answers.

1. SSP: $v'' = 0$, $v(0) = T_0$, $v(a) = T_0$, $v(x) = T_0$.

TP: $w_{xx} = (\frac{1}{k})w_t$, $0 < x < a$, $0 < t$; $w(0,t) = 0$, $w(a,t) = 0$, $w(x,0) = T_1 - T_0$.

EVP: $\phi'' + \lambda^2 \phi = 0$, $\phi(0) = 0$, $\phi(a) = 0$.

$$\phi_n(x) = \sin(\lambda_n x), \quad \lambda_n = \frac{n\pi}{a}, \quad n = 1, 2, \cdots .$$

$$w(x,t) = \sum b_n \sin(\lambda_n x) e^{-\lambda_n^2 kt},$$

$$b_n = \frac{2}{a} \int_0^a (T_1 - T_0) \sin(\lambda_n x) dx = (T_1 - T_0)\frac{2(1 - \cos(n\pi))}{n\pi}$$

3. SSP: $v'' + r = 0$, $v(0) = T_0$, $v(a) = T_0$, $v(x) = T_0 + \dfrac{rx(a - x)}{2}$.

TP: $w_{xx} = (\frac{1}{k})w_t$; $w(0,t) = 0$, $w(a,t) = 0$; $w(x,0) = T_1 - T_0 - \dfrac{rx(a - x)}{2}$.

EVP: $\phi'' + \lambda^2 \phi = 0$, $\phi(0) = 0$, $\phi(a) = 0$.

$$\phi_n(x) = \sin(\lambda_n x), \quad \lambda_n = \frac{n\pi}{a}, \quad n = 1, 2, \cdots$$

$$w(x,t) = \sum b_n \sin(\lambda_n x) e^{-\lambda_n^2 kt}$$

$$b_n = \frac{2}{a} \int_0^a \left(T_1 - T_0 - \frac{rx(a - x)}{2} \right) \sin(\lambda_n x) dx = 2(1 - \cos(n\pi)) \left[\frac{T_1 - T_0}{n\pi} - \frac{ra^2}{(n\pi)^3} \right].$$

5. SSP: $v'' - \gamma^2 v = 0$, $v'(0) = 0$, $v'(a) = 0$; $v(x) \equiv 0$, [so $w = u$].

TP: $u_{xx} - \gamma^2 u = (\frac{1}{k})u_t$; $u_x(0,t) = 0$, $u_x(a,t) = 0$; $u(x,0) = \dfrac{T_1 x}{a}$.

EVP: $\phi'' + \lambda^2 \phi = 0$, $\phi'(0) = 0$, $\phi'(a) = 0$, $\phi_0 = 1$, $\lambda_0 = 0$; $\phi_n(x) = \cos(\lambda_n x)$, $\lambda_n = n\pi/a$, $n = 1, 2, \cdots$.

$$u(x,t) = a_0 e^{-\gamma^2 t} + \sum a_n \cos(\lambda_n x) e^{-(\lambda_n^2 + \gamma^2)kt}$$

$$a_0 = \frac{T_1}{2}, \quad a_n = \frac{2}{a} \int_0^a \frac{T_1 x}{a} \cos(\lambda_n x) dx = \frac{2T_1}{\pi^2} \frac{(\cos(n\pi) - 1)}{n^2}.$$

7. SSP: $v'' = 0$, $v(0) = T_0$, $v(a) = T_0$; $v(x) = T_0$.

TP: $w_{xx} = (\frac{1}{k})w_t$; $w(0,t) = 0$, $w(a,t) = 0$; $w(x,0) = 0$. It should be obvious that $w(x,t) \equiv 0$, so that $u(x,t) = T_0$.

9. SSP: $v'' = 0$, $\quad v(0) = T_0$, $\quad v'(a) = 0$; $\quad v(x) = T_0$.

TP: $w_{xx} = (\frac{1}{k})w_t$; $\quad w(0,t) = 0$, $\quad w_x(a,t) = 0$; $\quad w(x,0) = T_1 - T_0$.

EVP: $\phi'' + \lambda^2\phi = 0$, $\quad \phi(0) = 0$, $\quad \phi'(a) = 0$.

$$\phi_n(x) = \sin(\lambda_n x), \quad \lambda_n = (n - \frac{1}{2})\pi/a, \quad n = 1, 2, \cdots.$$

$$w(x,t) = \sum c_n \sin(\lambda_n x)e^{-\lambda_n^2 kt},$$

$$c_n = \frac{2}{a}\int_0^a (T_1 - T_0)\sin(\lambda_n x)dx = \frac{2(T_1 - T_0)}{(n - \frac{1}{2})\pi}.$$

11. SSP: $v'' = 0$, $\quad v(0) = T_0$, v bounded. $v(x) = T_0$.

TP: $w_{xx} = (\frac{1}{k})w_t$, $\quad w(0,t) = 0$, $\quad w(x,0) = -T_0 e^{-\alpha x}$.

EVP: $\phi'' + \lambda^2\phi = 0$, $\quad \phi(0) = 0$, ϕ bounded. $\phi(x;\lambda) = \sin(\lambda x)$. $\lambda > 0$.

$$w(x,t) = \int_0^\infty B(\lambda)\sin(\lambda x)e^{-\lambda^2 kt}d\lambda,$$

$$B(\lambda) = \frac{2}{\pi}\int_0^\infty -T_0 e^{-\alpha x}\sin(\lambda x)dx = -\frac{2T_0}{\pi}\frac{\lambda}{\alpha^2 + \lambda^2}.$$

13. No need for SSP. $w(x,t) = u(x,t)$.

EVP: $\phi'' + \lambda^2\phi = 0$, $\quad \phi'(0) = 0$, ϕ bounded. $\phi(x;\lambda) = \cos(\lambda x)$, $0 \leq \lambda$.

$$u(x,t) = \int_0^\infty A(\lambda)\cos(\lambda x)e^{-\lambda^2 kt}d\lambda$$

$$A(\lambda) = \frac{2}{\pi}\int_0^a T_0\cos(\lambda x)dx = \frac{2T_0 \sin(\lambda a)}{\lambda\pi}$$

15. No SSP.

EVP: $\phi'' + \lambda^2\phi = 0$, $\quad \phi$ bounded. $\phi(x;\lambda) = \cos(\lambda x)$ or $\sin(\lambda x)$, $\quad \lambda \geq 0$.

$$u(x,t) = \int_0^\infty \left(A(\lambda)\cos(\lambda x) + B(\lambda)\sin(\lambda x)\right)e^{-\lambda^2 kt}d\lambda$$

$$A(\lambda) = \frac{1}{\pi}\int_0^a T_0\cos(\lambda x)dx = \frac{T_0 \sin(\lambda a)}{\lambda\pi}.$$

$$B(\lambda) = \frac{1}{\pi}\int_0^a T_0\sin(\lambda x)dx = \frac{T_0(1 - \cos(\lambda a))}{\lambda\pi}.$$

This problem would be neater if the origin were moved to $x = a/2$.

17. One interpretation: u is the temperature in an insulated bar. The left end is insulated, but heat is entering the right end at a rate of $\kappa S A$, where A is the cross-sectional area of the bar. The temperature is initially uniform at 0. Since energy is entering at $x = a$ but cannot exit anywhere, the temperature must increase with time.

19. $u_{2xx} = 2$, $u_{2t} = 2k$; $u_{3xx} = 6x$, $u_{3t} = 6kx$. Set $u = c_0 + c_1 x + c_2(x^2 + 2kt) + c_3(x^3 + 6kxt)$. Apply boundary conditions.

$$u(0,t) = 0: \quad c_0 + c_2 \cdot 2kt = 0, \quad 0 < t,$$

gives $c_0 = 0$, $c_2 = 0$. Then

$$u(a,t) = t: \quad c_1 a + c_3(a^3 + 6kat) = t$$

gives $c_3 = \dfrac{1}{6ka}$, $c_1 = -\dfrac{a}{6k}$. Finally, $u(x,t) = \dfrac{-a^2 x + x^3 + 6kxt}{6ka}$.

21. Set $w(x,t) = -\dfrac{2u_x}{u}$. Then $u_{xx} = u_t$; $u_x(0,t) = 0$, $u_x(1,t) = 0$; $\dfrac{u_x(x,0)}{u(x,0)} = -\dfrac{1}{2}$, $0 < x < a$. Because the initial condition is known as a function of x, we can solve it as if it were a differential equation. Then $u(x,0) = ce^{-x/2}$. The constant c is undetermined, but it will cancel out of w.

$$u(x,t) = a_0 + \sum_{1}^{\infty} a_n \cos(n\pi x)e^{-n^2\pi^2 t}$$

$$a_0 = 2c\left(1 - e^{-\frac{1}{2}}\right), \quad a_n = 2\int_0^1 ce^{-\frac{x}{2}}\cos(n\pi x)dx = c\frac{1 - e^{-\frac{1}{2}}\cos(n\pi)}{(n\pi)^2 + \frac{1}{4}}.$$

Finally, $w = \dfrac{-2u_x}{u}$.

23. If you know some linear algebra, use eigenvalues and eigenvectors. Otherwise, notice that the right-hand sides are identical except for sign; add the two equations to find

$$\frac{d}{dt}(c_1 u_1 + c_2 u_2) = 0.$$

Hence $c_1 u_1 + c_2 u_2$ is constant, and the initial condition means that constant is $c_1 T_0$. Now use $c_1 u_1 + c_2 u_2 = c_1 T_0$ to eliminate u_2 from the first equation. This simplifies to $\dot{u}_1 = \dfrac{h}{c_2} T_0 - \gamma u_1$ with $\gamma = h\left(\dfrac{1}{c_1} + \dfrac{1}{c_2}\right)$. Now solve this linear, first-order, nonhomogeneous equation:

$$u_1(t) = \frac{h}{c_2\gamma}T_0 + Ke^{-\gamma t}$$

with K an arbitrary constant. Apply the initial condition to find $K = \dfrac{h}{c_1\gamma}$.

25. Follow the hint (and see Section 0-4, Exercise 4). $u = \dfrac{v}{\rho}$; $u_\rho = \dfrac{(\rho v_\rho - v)}{\rho^2}$; $\rho^2 u_\rho = \rho v_\rho - v$; $(\rho^2 u_0 \rho)_\rho = \rho v_{\rho\rho} + v_\rho - v_\rho$. The partial differential equation becomes $\dfrac{v_{\rho\rho}}{\rho} = \dfrac{v_t}{k\rho}$ or $v_{\rho\rho} = \frac{1}{k}v_t$. From the left boundary condition, $v(0,t) = 0$ (otherwise v/ρ can't be bounded as $\rho \to 0$). And $v(a,t) = 0$. The initial condition is $v(\rho,0) = T_0\rho$. The problem for v is now routine (see Section 2-3 Example).

$$v(x,t) = \sum b_n \sin(\lambda_n \rho)e^{-\lambda_n^2 kt},$$

$$\lambda_n = \frac{n\pi}{a}, \quad b_n = \frac{2}{a}\int_0^a T_0 \rho \sin(\lambda_n \rho)d\rho = \frac{-2T_0 a \cos(n\pi)}{n\pi}.$$

27. The steady-state problem (differential equation in standard form) is

$$v'' - \gamma^2 v = -\gamma^2 (T_0 + Sx)$$

$$v(0) = T_0, \quad v'(a) = 0$$

The general solution of the differential equation is

$$v(x) = T_0 + Sx + c_1 \cosh \gamma x + c_2 \sinh(\gamma x)$$

Apply the boundary conditions:

$$v(0) = T_0: \quad T_0 + c_1 = T_0 \quad (c_1 = 0)$$

$$v'(a) = 0: \quad S + c_2 \gamma \cosh(\gamma a) = 0 \quad \left(c_2 = \frac{-S}{\gamma \cosh(\gamma a)} \right).$$

29. Set $\lambda = 0$. The differential equation becomes $\phi'' = 0$, with general solution $\phi(x) = c_1 + c_2 x$. Since $\phi'(x) = c_2$, both boundary conditions require $c_2 = 0$, leaving c_1 arbitrary. Hence $\lambda = 0$ is an eigenvalue, $\phi_0(x) = 1$ is an eigenfunction.

31. In the given formula, set $x = 0$ so that the boundary condition becomes

$$\int_0^\infty B(\omega) \sin(\omega t) d\omega = f(t), \quad 0 < t$$

This is a Fourier sine integral problem, so we choose

$$B(\omega) = \frac{2}{\pi} \int_0^\infty f(t) \sin(\omega t) dt.$$

33.a. With $I = $ const., $\phi \equiv 0$, and $h(x, t) \to v(x)$, the problem becomes $Tv'' + aKv' = -I$, $v(0) = h_1$, $v(L) = h_2$. The general solution is $v(x) = -\dfrac{Ix}{aK} + c_1 + c_2 e^{-\gamma x}$, with $\gamma = \dfrac{aK}{T}$. Apply the boundary conditions.

$$v(0) = h_1: \quad c_1 + c_2 = h_1$$

$$v(L) = h_2: \quad \frac{-IL}{aK} + c_1 + c_2 e^{-\gamma L} = h_2.$$

[Note that c_1, c_2 in Answers means something different.] From here

$$c_1 = h_1 - c_2, \quad c_2 = \frac{\left(h_2 - h_1 + \frac{IL}{aK} \right)}{(e^{-\gamma L} - 1)}$$

b. See Answers.

c. See Answers. Since $h_0(x)$ is not given explicitly, we have only a formula for

$$c_n = \frac{2}{a} \int_0^a e^{\frac{\mu x}{2}} (h_0(x) - v(x)) \sin \left(\frac{n\pi x}{L} \right) dx.$$

35. The boundary conditions are:

$$\frac{\partial u}{\partial x}(0,t) = \frac{\partial S}{\partial x}(0,t) + \frac{\partial P}{\partial x}(0,t) = 0$$

$$u(L,t) = S(L,t) + P(L,t) = S_0.$$

The initial condition is $u(x,0) = S(x,0) + P(x,0) = 0$. The problem for u is just the same as Exercise 11, with S_0 replacing T_0, 0 replacing T_1 and D replacing k. Hence

$$u(x,t) = S_0 - \sum_{n=1}^{\infty} \frac{2S_0(-1)^{n+1}}{(n-\frac{1}{2})\pi} \cos(\lambda_n x) \exp(-\lambda_n^2 Dt)$$

37. a. Use odd and even properties for the integrations.

$$\int_{-c}^{c} \alpha T E \, dy = \alpha E \cdot 2 \int_0^c T_0 \, dy = \alpha E T_0 (2c)$$

$$\int_{-c}^{c} \alpha T E y \, dy = \alpha E \cdot 2 \int_0^c S y^2 \, dy = \alpha E S \frac{2c^3}{3}$$

b. The problem for T is

$$\frac{\partial^2 T}{\partial y^2} = \frac{1}{k}\frac{\partial T}{\partial t}, \quad -c < y < c, \;\; 0 < t$$

$$T(c,t) = 200, \quad T(-c,t) = 500, \quad 0 < t$$

$$T(y,0) = 500.$$

Steady state solution: $v(y) = -150 + 350y/c$.

Transient problem:

$$\frac{\partial^2 w}{\partial y^2} = \frac{1}{k}\frac{\partial w}{\partial t}, \quad -c < y < c, \;\; 0 < t$$

$$w(\pm c, t) = 0, \quad 0 < t$$

$$w(y,0) = 500 - v(y) = 350(1 - y/c)$$

[EVP:] $\phi'' + \lambda^2 \phi = 0, \quad \phi(\pm c) = 0,$

Solution: $\lambda_n = n\pi/2c, \; \phi_n(y) = \sin(\lambda_n(y+c))$.

Solution of partial differential equation and boundary condition:

$$w(y,t) = \sum_{n=1}^{\infty} b_n \sin(\lambda_n(y+c)) e^{-\lambda_n^2 kt}$$

$$b_n = \frac{-2}{2c} \int_{-c}^{c} 350(1-y/c)\sin(\lambda_n(y+c))\,dy = \frac{400\cos(n\pi) + 1000}{n\pi}$$

The integration is easy with the substitutions $z = y + c$.

c. See Answers.

Chapter 3

3.1 The Vibrating String

1. The unknown u is displacement, so $[u] = L$. Then $[u_{xx}] = 1/L$, $[u_{tt}] = L/t^2$. $c^2 = T/\rho$, so

$$[c^2] = \frac{F}{m/L} = \frac{mL/t^2}{m/L} = \left(\frac{L}{t}\right)^2, \left[\frac{g}{c^2}\right] = \frac{L/t^2}{L^2/t^2} = \frac{1}{L}.$$

Then each term in Equation (5) has dimension $\frac{1}{L}$.

3. Let the time-independent solution be $v(x)$. Then Equations (5) and (8) become $v'' = g/c^2$, $v(0) = 0$, $v(a) = 0$, with solution $v(x) = x(x-a)g/2c^2$. The graph is a parabola opening up. It lies below the x-axis for $0 < x < a$.

Chapter 3

3.2 Solution of the Vibrating String Problem

1. Compute the second derivatives:

$$\frac{\partial^2 u_n}{\partial x^2} = -\lambda_n^2 u_n, \quad \frac{\partial^2 u_n}{\partial t^2} = -\lambda_n^2 c^2 u_n.$$

Thus, u_n satisfies the wave equation. For the boundary conditions, substitute for x:

$x = 0 : \sin(0) = 0$ making $u_n(0, t) = 0$

$x = a : \sin(\lambda_n a) = \sin(n\pi) = 0$, making $u_n(a, t) = 0$.

3. Use the solution Equation (8) and coefficients found from Equation (9) and (10).

$$a_n = 0, \quad b_n = \frac{2}{n\pi c} \int_0^a \sin(\frac{n\pi x}{a}) dx = \frac{2a(1 - \cos(n\pi))}{(n\pi)^2 c}.$$

5.

$$b_n = 0, \quad a_n = \frac{2}{a} \int_0^{\frac{a}{2}} U_0 \sin(\frac{n\pi x}{a}) dx = \frac{2U_0(1 - \cos(n\pi/2))}{n\pi}$$

7. The equilibrium solution is $v(x) = p_0$. Let $p(x, t) = p_0 + w(x, t)$ (thus, w is gauge pressure) and then $w_{xx} = (\frac{1}{c^2}) w_{tt}$.

a. The boundary conditions are $w(0, t) = 0$, $w(a, t) = 0$. EVP: $\phi'' + \lambda^2 \phi = 0$, $\phi(0) = 0$, $\phi(a) = 0$. Solution $\phi_n(x) = \sin(\lambda_n x)$, $\lambda_n = \frac{n\pi}{a}$, $n = 1, 2, \cdots$.

b. The boundary conditions are $w(0, t) = 0$, $w_x(a, t) = 0$. EVP: $\phi'' + \lambda^2 = 0$, $\phi(0) = 0$, $\phi'(a) = 0$. $\phi_n(x) = \sin(\lambda_n x)$, $\lambda_n = (n - \frac{1}{2})\pi/a$, $n = 1, 2, \cdots$.

9. Since the equation and boundary conditions are homogeneous, assume $u(x, t) = \phi(x)T(t)$. The partial differential equations separate as

$$\frac{\phi''(x)}{\phi(x)} = \frac{\ddot{T}(t)}{c^2 T(t)} + k\frac{\dot{T}(t)}{T(t)} = -\lambda^2.$$

EVP: $\phi'' + \lambda^2 \phi = 0$, $\phi(0) = 0$, $\phi(a) = 0$, $\phi_n(x) = \sin(\lambda_n x)$, $\lambda_n = n\pi/a$, $n = 1, 2, \cdots$.

Time problem: $\ddot{T} + c^2 k\dot{T} + c^2 \lambda^2 T = 0$. Solution $T(t) = e^{-\alpha t}(c_1 \cos(\beta t) + c_2 \sin(\beta t))$ with $\alpha = \frac{c^2 k}{2}$, $\beta = \sqrt{\lambda_n^2 c^2 - \alpha^2}$.

11. The partial differential equation and b.c. are all homogeneous, so we seek solutions $u(x, t) = \phi(x)T(t)$. The partial differential equation separates to

$$\frac{\phi''''}{\phi} = -\frac{\ddot{T}}{c^2 T} = +\lambda^4 \text{ (assuming a plus sign for the constant, because sines and cosines}$$
are expected for $T(t)$).

E.V.P.: $\phi'''' - \lambda^4 \phi = 0$, $\phi(0) = 0$, $\phi''(0) = 0$, $\phi(a) = 0$, $\phi''(a) = 0$. The general solution of the differential equation is $\phi(x) = c_1 e^{\lambda x} + c_2 e^{-\lambda x} + c_3 \cos(\lambda x) + c_4 \sin(\lambda x)$. Apply the b.c.

$$c_1 + c_2 + c_3 = 0$$

$$\lambda^2(c_1 + c_2 - c_3) = 0$$

$$c_1 e^{\lambda a} + c_2 e^{-\lambda a} + c_3 \cos(\lambda a) + c_4 \sin(\lambda a) = 0$$

$$\lambda^2 \left(c_1 e^{\lambda a} + c_2 e^{-\lambda a} - c_3 \cos(\lambda a) - c_4 \sin(\lambda a) \right) = 0$$

This is a system of 4 linear equations in 4 unknowns. Since it is a homogeneous system, its determinant must be 0 in order to get any non-zero solution. Another approach: note from the first two equations that $c_1 + c_2 = 0$ and $c_3 = 0$. Then the other two simplify (using $c_2 = -c_1$) to:

$$2c_1 \sinh(\lambda a) + c_4 \sin(\lambda a) = 0$$

$$2c_1 \sinh(\lambda a) - c_4 \sin(\lambda a) = 0$$

From these, $2c_1 \sinh(\lambda a) = 0$, making $c_1 = 0$; and $c_4 \sin(\lambda a) = 0$, making $c_4 = 0$ (and $\phi(x) \equiv 0$!) or else $\sin(\lambda a) = 0$.

Therefore $\phi_n(x) = \sin(\lambda_n x)$, $\lambda_n = \frac{n\pi}{a}$, $n = 1, 2, \cdots$. The solution of the problem for T is

$$T(t) = a_n \cos(\lambda_n^2 ct) + b_n \sin(\lambda_n^2 ct).$$

The frequencies are $\lambda_n^2 c$ rad/sec or $\lambda_n^2 c / 2\pi$ Hz.

13. The general solution of the differential equation for φ is

$$\varphi(x) = A \cos(\lambda x) + B \sin(\lambda x) + C \cosh(\lambda x) + D \sinh(\lambda x).$$

The boundary conditions at $x = 0$ are

$$\varphi(0) = 0: \qquad A + C = 0$$
$$\varphi'(0) = 0: \quad \lambda(B + D) = 0$$

Use these to solve for, say, A and B, so that

$$\varphi(x) = C \left(\cosh(\lambda x) - \cos(\lambda x) \right) + D \left(\sinh(\lambda x) - \sin(\lambda x) \right).$$

Now apply the boundary conditions at $x = a$. After dividing by λ^2 and λ^3, these become

$$\phi''(a) = 0: \quad C \left(\cosh(\lambda a) + \cos(\lambda a) \right) + D \left(\sinh(\lambda a) + \sin(\lambda a) \right) = 0$$

$$\phi'''(a) = 0: \quad C \left(\sinh(\lambda a) - \sin(\lambda a) \right) + D \left(\cosh(\lambda a) + \cos(\lambda a) \right) = 0$$

This is a system of two homogeneous, linear, algebraic equations in the two unknowns C and D. There is a nonzero solution if and only if its determinant is 0:

$$[\cosh(\lambda a) + \cos(\lambda a)]^2 - \left[\sinh^2(\lambda a) - \sin^2(\lambda a) \right] = 0.$$

Expand and use the identities $\sin^2 \theta + \cos^2 \theta = 1$ and $\cosh^2 \theta - \sinh^2 \theta = 1$ to reduce this equation to

$$1 + \cos(\lambda a) \cosh(\lambda a) = 0 \quad \text{or} \quad \cos(\lambda a) = -\frac{1}{\cosh(\lambda a)}. \tag{1}$$

Since the hyperbolic cosine is never less than 1, the $\cos(\lambda a)$ has to be negative. A sketch of $\cos(\theta)$ and $-1/\cosh(\theta)$ on the same axes shows that there are two solutions of Equation (1) in the range $\frac{\pi}{2} < \lambda a < \frac{3\pi}{2}$. These solutions are approximately $\lambda a = 1.875$ and $\lambda a = 4.693$. From here, we can determine λ_1 and λ_2.

Now, the coefficients C and D must satisfy the equation

$$\frac{C}{D} = -\frac{\cosh(\lambda a) + \cos(\lambda a)}{\sinh(\lambda a) - \sin(\lambda a)}.$$

This ratio is approximately -1.362 and -0.982 for the two values of λ. Graphs of the eigenfunctions φ_1 and φ_2 are shown below.

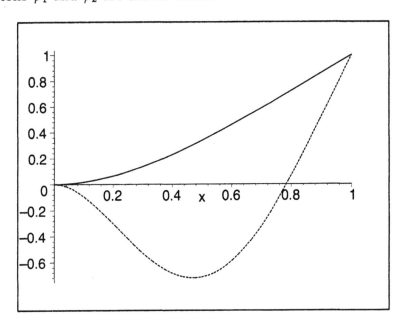

Exercise 11

15. Separated partial differential equations

$$\frac{\phi''}{\phi} = \frac{T''}{c^2 T} + \gamma^2 = -\lambda^2$$

EVP: $\phi'' + \lambda^2 \phi = 0, \quad \phi(0) = 0, \quad \phi(a) = 0.$

Solution: $\phi_n(x) = \sin(\lambda_n x), \quad \lambda_n = \frac{n\pi}{a}, \quad n = 1, 2, \cdots.$

$$T'' + (\lambda^2 + \gamma^2)c^2 T = 0, \quad T(t) = a_n \cos(\mu_n t) + b_n \sin(\mu_n t), \quad \mu_n = \sqrt{\lambda_n^2 + \gamma^2}\, c.$$

$$u(x, t) = \sum_{n=1}^{\infty} (a_n \cos(\mu_n t) + b_n \sin(\mu_n t)) \sin(\lambda_n x).$$

Apply the initial conditions:

$$u(x, 0) = h: \quad \sum_1^{\infty} a_n \sin(\lambda_n x) = h, \quad 0 < x < a;$$

$$a_n = \frac{2h(1 - \cos(n\pi))}{n\pi}.$$

$$\sum b_n \mu_n \sin(\lambda_n x) = 0 : \quad b_n = 0.$$

17. The general coefficient in the series in $\dfrac{\sin(n\pi)}{n^2}$. Since the absolute value of this coefficient is $\leq 1/n^2$, and $\sum 1/n^2$ converges, the series in Equation (15) does converge uniformly.

Chapter 3

3.3 D'Alembert's Solution

1. Since $g(x) \equiv 0$, $u(x,t) = \frac{1}{2}\left[\bar{f}_0(x+ct) + \bar{f}_0(x-ct)\right]$. It is useful to sketch $\bar{f}_0$ (see Sec. 3-2, Figure 3). Whenever convenient, use oddness and periodicity to simplify calculations. For example $\bar{f}_0(1.65a) = \bar{f}_0(-.35a) = -\bar{f}_0(.35a) = -.7h$.

$$u\left(.25a, \frac{.2a}{c}\right) = \frac{1}{2}\left[\bar{f}_0(.45a) + \bar{f}_0(.05a)\right] = \frac{1}{2}[.9h + .1h] = .5h$$

$$u\left(.25a, \frac{.4a}{c}\right) = \frac{1}{2}\left[\bar{f}_0(.65a) + \bar{f}_0(-.15a)\right] = \frac{1}{2}[.7h - .3h] = .2h$$

$$u\left(.25a, \frac{.8a}{c}\right) = \frac{1}{2}\left[\bar{f}_0(1.05a) + \bar{f}_0(-.55a)\right] = \frac{1}{2}[-.1h + (-.9h)] = -.5h$$

$$u\left(.25a, \frac{1.4a}{c}\right) = \frac{1}{2}\left[\bar{f}_0(1.65a) + \bar{f}_0(-1.15a)\right] = \frac{1}{2}[-.7h + .3h] = -.2h$$

3. Since $f(x) \equiv 0$, $u(x,t) = \frac{1}{2}\left[\bar{G}_e(x+ct) - \bar{G}_e(x-ct)\right]$. For $g(x) = ac$, $0 < x < a$,

$$G(x) = \frac{1}{c}\int_0^x g(x')dx' = \alpha x, \quad 0 < x < a.$$ It is useful to sketch $\bar{G}_e(x)$. (See Figure 10 in Section 1-4.)

Whenever convenient, use evenness and periodicity to simplify. For instance, $\bar{G}_e(1.7a) = \bar{G}_e(-.3a) = \bar{G}_e(.3a) = .3\alpha a$.

$$u\left(0, \frac{.5a}{c}\right) = \frac{1}{2}\left[\bar{G}_e(.5a) - \bar{G}_e(-.5a)\right] = 0$$

(of course!)

$$u\left(.2a, \frac{.6a}{c}\right) = \frac{1}{2}\left[\bar{G}_e(.8a) - \bar{G}_e(-.4a)\right] = \frac{1}{2}[.8\alpha a - .4\alpha a] = .2\alpha a$$

$$u\left(.5a, \frac{1.2a}{c}\right) = \frac{1}{2}\left[\bar{G}_e(1.7a) - \bar{G}_e(-.7a)\right] = \frac{1}{2}[.3\alpha a - .7\alpha a] = -.2\alpha a$$

5. $G(x) = \frac{1}{c}\int_0^x g(x')dx'$. To carry out the integration, distinguish three cases.

(a) $0 < x < .4a$; then also $0 < x' < .4a$, so $G(x) = 0$.

(b) $.4a < x < .6a$. The integration has to be broken at $.4a$:

$$G(x) = \int_0^{.4a} 0dx' + \int_{.4a}^x 5dx' = 5(x - .4a)$$

(c) $.6a < x < a$. The integral is broken at two places:

$$G(x) = \int_0^{.4a} 0dx' + \int_{.4a}^{.6a} 5dx' + \int_{.6a}^x 0dx' = (.2a) \cdot 5 = a.$$

Now put the pieces together,

$$G(x) = \begin{cases} 0, & 0 < x < .4a \\ 5(x - .4a), & .4a < x < .6a \\ a, & .6a < x < a \end{cases}$$

7. See Answers.

9. See Answers.

11. See Answers.

13. Look for a solution of Equation (*) that is a function of t alone – say $v(t)$. Then $v'' = c^2 \cos(t)$, so $v(t) = -c^2 \cos(t)$ (by two integrations). Then $u(x, t) = v(t) + w(x, t)$, where w is a general solution of the wave equation, $w_{xx} = \dfrac{w_{tt}}{c^2}$. We may use Equation (1) to represent w.

Chapter 3

3.4 One-Dimensional Wave Equations: Generalities

1. First, see Section 2-7, for a discussion of expansion in series of eigenfunction. Using Equation (15) for w, the two initial conditions become just such series:

$$\sum_{n=1}^{\infty} a_n \phi_n(x) = f(x) - v(x), \quad l < x < r$$

$$\sum_{n=1}^{\infty} (b_n \lambda_n c) \phi_n(x) = g(x), \quad l < x < r$$

In the first equation, a_n is determined by orthogonality to be as given in Equation (16). In the second equation, the quantity in parentheses is the coefficient. Hence

$$b_n \lambda_n c = \int_{l}^{r} g(x) \phi_n(x) p(x) dx / I_n$$

From here, find Equation (17).

3. $T_n(t) = a_n \cos(\lambda_n ct) + b_n \sin(\lambda_n ct)$. The sine and cosine functions have period 2π in their argument, hence period $\dfrac{2\pi}{\lambda_n c}$ (dimensions of time–usually seconds) and frequency $\dfrac{\lambda_n c}{2\pi}$ (usually Hz).

5. See Answers.

7. See Answers.

Chapter 3

3.5 Estimation of Eigenvalues

1. See Answers.

3. Try $y = \sin(\pi x)$ to satisfy b.c.

$$\int_0^1 s(y')^2 dx = \int_0^1 \pi^2 \cos^2(\pi x) dx = \frac{\pi^2}{2}$$

$$\int_0^1 py^2 dx = \int_0^1 (1+x)\sin^2(\pi x) dx = \int_0^1 (1+x)(1 - \cos(2\pi x))\frac{dx}{2}$$

$$= \frac{1}{2}\left[x + \frac{x^2}{2} - \frac{\sin(2\pi x)}{2\pi} - \frac{\cos(2\pi x)}{(2\pi)^2} - \frac{x\sin(2\pi x)}{2\pi} \right]\Big|_0^1 = \frac{3}{4}.$$

Note that $q(x) \equiv 0$. The ratio of Equation (4) is

$$\lambda_1^2 \le \frac{\pi^2/2}{3/4} = \frac{2\pi^2}{3} \cong 6.58.$$

Another function that satisfies the b.c. is $y = x - x^2$. The (easy) integrals are

$$\int_0^1 s(y')^2 dx = \int_0^1 (1 - 2x)^2 dx = \frac{1}{3}$$

$$\int_0^1 (1+x)(x - x^2)^2 dx = \frac{1}{20}$$

Then Equation (4) gives

$$\lambda_1^2 \le \frac{1/3}{1/20} = 6.66.$$

5. Note that $s(x) = 1$, $\quad p(x) = \frac{1}{x^4}$, $q(x) \equiv 0$. The integrals are

$$\int_1^2 (y')^2 dx = \int_1^2 (4x^2 - 12x + 9) dx = \frac{1}{3}$$

$$\int_1^2 \frac{y^2}{x^4} dx = \int_1^2 (x^4 - 6x^3 + 13x^2 - 12x + 4)\frac{dx}{x^4} = x - 6\ln x - 13x^{-1} + 6x^{-2} - \frac{4x^{-3}}{3}\Big|_1^2$$

$$= \frac{25}{6} - 6\ln(2).$$

Then Equation (4) gives

$$\lambda_1^2 \le \frac{\frac{1}{3}}{\frac{25}{6} - 6\ln(2)} \cong 42.8.$$

The true value is $\lambda_1^2 = (2\pi)^2 \cong 39.478$.

Chapter 3

3.6 Wave Equation in Unbounded Regions

1. The derivation of ψ and ϕ follows the text through the application of the initial conditions. The boundary condition at $x = 0$ gives

$$\bar{f}'(ct) + \bar{G}'(ct) + \bar{f}'(-ct) - \bar{G}'(-ct) = 0.$$

From here, $\bar{f}'$ must be odd and $\bar{G}'$ must be even. Hence $\bar{f}$ is even and $\bar{G}$ is odd, giving the formula in the Answers.

3. See Answers.

5. See Answers.

7. The derivation of ψ and ϕ follows the text through application of the initial conditions. However, the initial conditions are valid for all x; hence ψ and ϕ are completely determined - no extensions are necessary (or possible!). The formula for the solution can be written as in Exercise 8 or as

$$u(x,t) = \frac{1}{2}(f(x + ct) + f(x - ct)) + \frac{1}{2}(G(x + ct) - G(x - ct))$$

with $G(x) = \dfrac{1}{c}\displaystyle\int_0^x g(y)dy$, for all x.

Chapter 3

Miscellaneous

1. Use the solution in Section 3-2, Equations (8)-(10). It is necessary only to determine the coefficients. Because $g(x) \equiv 0$, $b_n = 0$, $a_n = \dfrac{2}{a} \displaystyle\int_0^a 1 \cdot \sin\left(\dfrac{n\pi x}{a}\right) dx = \dfrac{2(1 - \cos(n\pi))}{n\pi}$.

3. Use the D'Alembert solution of Section 3.3. Since $g(x) \equiv 0$, $u(x,t) = \dfrac{1}{2}\left[\bar{f}_0(x + ct) + \bar{f}_0(x - ct)\right]$.

   Here, $\bar{f}_0$ is the odd square wave, which can take only the values $1, -1$ (and 0, but only at isolated points, so we'll ignore that). Now see Answers.

5. See Answers.

7. See Answers.

9. See Answers.

11. From Section 3-6, the solution is $u(x,t) = \phi(x - ct)$, where

$$\phi(q) = \begin{cases} 0, & q > 0 \\ \sin(\frac{q}{a}), & 0 < q < \pi a \\ 0, & \pi a < q \end{cases}$$

   In the Answers, see a sketch of ϕ. Now the graph of $u(x,t) = \phi(x - ct)$ vs. x is simply the graph of ϕ shifted ct units to the right.

13. See Answers and the solution of 11 just above.

15. Use $y(x) = x(1 - x)$, so $y' = 1 - 2x$. The required integrals all have polynomial integrands:

$$\int_0^1 y'^2 dx = \int_0^1 (1 - 4x + 4x^2)dx = \frac{1}{3}$$

$$\int_0^1 y^2 dx = \int_0^1 x^2(1 - 2x + x^2)dx = \frac{1}{30}$$

$$\int_0^1 xy^2 dx = \int_0^1 x^3(1 - 2x + x^2)dx = \frac{1}{60}$$

   From Equation (4), $\lambda_1^2 \leq 10.5$.

17. $12a^2\text{sech}^2(ax - 4a^3 t) = f(x - ct)$.

a. $x - 4a^2 t = x - ct$, so $c = 4a^2$.

b. $f(q) = 12a^2\text{sech}^2(aq)$.

19. With v defined as shown, Equation (a) is

$$\frac{\partial^2 v}{\partial t \partial x} = \frac{\partial^2 v}{\partial x \partial t}$$

which is true if the derivatives are continuous. Then Equation (b) is

$$-\frac{\partial^2 v}{\partial t^2} = -c^2 \frac{\partial^2 v}{\partial x^2}$$

21. The restated problem is

$$\frac{\partial^2 v}{\partial x^2} = \frac{1}{c^2} \frac{\partial^2 v}{\partial t^2}, \quad 0 < x < a, \quad 0 < t$$

$$v(0,t) = 0, \quad v_x(a,t) = 0, \quad 0 < t$$

$$v(x,0) = U_0 x, \quad v_t(x,0) = 0, \quad 0 < x < a.$$

Solution by separation of variables is now routine. See Answers.

23. Using the given form, find

$$u_t = \psi'(x+Vt)V\phi(x-V_t)Y(y) - \psi(x+Vt)\phi'(x-Vt)VY(y)$$

$$u_x = \psi'(x+Vt)\phi(x-Vt)Y(y) + \psi(x+Vt)\phi'(x-Vt)Y(y)$$

$$u_t + Vu_x = 2V\psi'(x+Vt)\phi(x-Vt)Y(y)$$

$$u_{yy} = \psi(x+Vt)\phi(x-Vt)Y''(y)$$

Matching sides of the equation, we see that $\phi(x-Vt)$ drops out. The remainder is equivalent to

$$\frac{2V\psi'}{k\psi} = \frac{Y''}{Y} = \text{constant.}$$

25. $u(x,y,t) = \displaystyle\sum_{n=1}^{\infty} \phi_n(x-Vt)\exp(-\mu_n(x+Vt))\sin(\lambda_n y)$ where $\mu_n = \dfrac{\lambda_n^2 k}{2V}$. The two conditions are

$$u(x,y,0) = T_0 : \sum_{n=1}^{\infty} \phi_n(x)e^{-\mu_n x}\sin(\lambda_n y) = T_0, \quad 0 < y < b$$

$$u(0,y,t) = T_1 : \sum_{n=1}^{\infty} \phi_n(-Vt)e^{-\mu_n Vt}\sin(\lambda_n y) = T_1, \quad 0 < y < b$$

To satisfy both equations we must have

$$\phi_n(x)e^{-\mu_n x} = \frac{2}{b}\int_0^b T_0 \sin(\lambda_n y)\,dy$$

$$\phi_n(-Vt)e^{-\mu_n Vt} = \frac{2}{b}\int_0^b T_1 \sin(\lambda_n y)\,dy.$$

The first equation describes $\phi_n(q)$ for positive arguments, the second for negative arguments.

27. From 26, $\phi'' = \dfrac{-c\phi'}{k}$ so that $\phi(x - ct) = \exp\left(\dfrac{-c(x - ct)}{k}\right)$. Now take c as required, so

that $c^2 = 2i\dfrac{\omega k}{2} = i\omega k$, and $\dfrac{c^2 t - cs}{k} = i\omega t - (1 + i)px$. Next,

$$\phi(x - ct) = \exp\left(\frac{-c(x - ct)}{k}\right) = \exp\left(-px + i(\omega t - px)\right)$$

$$= e^{-px}\left(\cos(\omega t - ps) + i\sin(\omega t - px)\right)$$

and

$$\phi(x - \bar{c}t) = e^{-px}\left(\cos(\omega t - px) - i\sin(\omega t - px)\right).$$

29. Do a simple differentiation:

$$u_{xx} = -\left(\frac{\omega}{c}\right)^2 u; \quad u_{tt} = -\omega^2 u.$$

The b.c. work by direct evaluation.

31. Using the product form for u, the partial differential equation becomes

$$\phi^{(2)}T - \epsilon\phi^{(4)}T = \frac{1}{c^2}\phi T''$$

and, after division by ϕT,

$$\frac{\phi^{(2)} - \epsilon\phi^{(4)}}{\phi} = \frac{T''}{c^2 T} = -\lambda^2$$

The boundary conditions separate in the usual way, so the eigenvalue problem is

$$\phi^{(2)} - \epsilon\phi^{(4)} + \lambda^2\phi = 0$$

$$\phi(0) = 0, \quad \phi(a) = 0$$

$$\phi''(0) = 0, \quad \phi''(a) = 0$$

33. From Exercise 32, $\mu_n = n\pi/a$ and

$$\lambda^2 = \mu^2 + \epsilon\mu^4$$

For any fixed n, if ϵ is small enough

$$\lambda_n = \frac{n\pi}{a}\sqrt{1 + \epsilon\left(\frac{n\pi}{a}\right)^2} \cong \frac{n\pi}{a}$$

Chapter 4

4.1 Potential Equation

1. $p_{xx} = 2d$, $p_{yy} = 2f$. The potential equation becomes $2(d + f) = 0$. Thus, if $f = -d$, the given polynomial is a solution for any choice of a, b, c, d, e.

Take for example $p(x,y) = x^2 - y^2 + xy$. Then $p_x = 2x + y$, $p_y = -2y + x$. Both of these are 0 only where $x = y = 0$. At this point, the second derivatives are of opposite signs and the determinant

$$\begin{vmatrix} p_{xx} & p_{xy} \\ p_{yx} & p_{yy} \end{vmatrix} = -5.$$

(See a calculus text.)

3. For the given function u, $\nabla^2 u = (Y'' - \pi^2 Y)\sin(\pi x)$. In order for u to be a solution of the potential equation, this must be 0, so Y must satisfy $Y'' - \pi^2 Y = 0$. Thus $Y(y) = c_1 \cosh(\pi y) + c_2 \sinh(\pi y)$. If we want $u(x, 0) = 0$, we must have $c_1 = 0$. Then $u(x, 1) = c_2 \sinh(\pi)\sin(\pi x) = \sin(\pi x)$ requires $c_1 = \dfrac{1}{\sinh(\pi)}$.

5. If v is a function of r alone, the potential equation reduces to $\dfrac{1}{r}(rv')' = 0$, which we solved in Section 0-1, Exercise 11: $v(r) = c_1 + c_2 \ln(r)$.

7. Apply the chain rule; for instance

$$\frac{\partial u}{\partial x} = \frac{\partial v}{\partial r}\frac{\partial r}{\partial x} + \frac{\partial v}{\partial \theta}\frac{\partial \theta}{\partial x}$$

Now, $r^2 = x^2 + y^2$ so $2rr_x = 2x$ or $r_x = \dfrac{x}{r}$. Since $x = r\cos(\theta)$, $r_x = \cos(\theta)$. Also $\theta = \tan^{-1}\left(\dfrac{y}{x}\right)$ so $\theta_x = \dfrac{-\frac{y}{x^2}}{1 + \left(\frac{y}{x}\right)^2} = -\dfrac{y}{(x^2 + y^2)}$. Since $y = r\sin(\theta)$, $\theta_x = -\dfrac{r\sin(\theta)}{r^2} = -\dfrac{\sin(\theta)}{r}$.

9. See Answers.

Chapter 4

4.2 Potential in a Rectangle

1. First note that $\sinh(\lambda(b-y))$ is really a solution – either directly or by using the hyperbolic identities (see Mathematical References) to express it as a combination of $\sinh(\lambda y)$ and $\cosh(\lambda y)$. Now, the determinant in Answers is just the Wronskian. (See Equation (7) of Section 0-1.) By direct calculation, it is $-\lambda \sinh(\lambda y)\cosh(\lambda(b-y)) - \lambda\cosh(\lambda y)\sinh(\lambda(b-y))$. Again referring to the Mathematical References, convert this to the quantity shown.

3. The center of the rectangle is $\left(\dfrac{a}{2}, \dfrac{b}{2}\right)$. The cosh in the numerator inside the series has value 1 there, and the sine becomes $\sin\left(\dfrac{n\pi}{2}\right)$. Thus

$$u\left(\frac{a}{2}, \frac{b}{2}\right) = \frac{8}{\pi^2}\sum_{n=1}^{\infty} \frac{\sin^2(n\pi/2)}{n^2 \cosh(n\pi b/2a)}$$

$$= \frac{8}{\pi^2}\left(\frac{1}{\cosh(\pi b/2a)} + \frac{1}{9\cosh(3\pi b/2a)} + \frac{1}{25\cosh(5\pi b/2a)} + \cdots\right)$$

Now, cosh increases very rapidly with its argument – see the table of $\cosh\left(\dfrac{n\pi b}{2a}\right)$ for odd n and three values of $\dfrac{b}{a}$.

b/a :	n :	1	3	5	7
0.5		1.325	5.323	25.387	122.078
1		2.509	55.663	1287.99	29804.87
2		11.592	6195.82	3317812	1.78×10^9

Here is a sample calculation for $\dfrac{b}{a} = \dfrac{1}{2}$.

$$u\left(\frac{a}{2}, \frac{b}{2}\right) \cong \frac{8}{\pi^2}\left[\frac{1}{1.325} + \frac{1}{9 \times 5.323} + \frac{1}{25 \times 25.387}\right] =$$

$$= \frac{8}{\pi^2}[.755 + .021 + .002] = .630$$

In the other two cases, 2 terms or 1 term is enough for three digits accuracy. For $b = 2a$, $u\left(\dfrac{a}{2}, \dfrac{b}{2}\right) \cong .070$.

5. Refer to the solution formula, Equation (9). Apply the boundary conditions at $y = 0, b$

$$u(x,0) = 0 : \quad \sum a_n \sin\left(\frac{n\pi x}{a}\right) = 0$$

from which $a_n = 0$.

$$u(x, b) = f(x): \quad \sum_{n=1}^{\infty} b_n \sinh\left(\frac{n\pi b}{a}\right) \sin\left(\frac{n\pi x}{a}\right) = f(x), \quad 0 < x < a$$

from which

$$b_n \sinh\left(\frac{n\pi b}{a}\right) = \frac{2}{a} \int_0^a f(x) \sin\left(\frac{n\pi x}{a}\right) dx.$$

The integration has been done many times for the given $f(x)$:

$$b_n = \frac{8 \sin(n\pi/2)}{\pi^2 n^2 \sinh(n\pi b/a)}$$

Therefore

$$u(x, y) = \sum_{n=1}^{\infty} \frac{8 \sin(n\pi/2)}{n^2 \pi^2} \frac{\sinh(n\pi y/a)}{\sinh(n\pi b/a)} \sin\left(\frac{n\pi x}{a}\right).$$

See Figures for the surfaces.

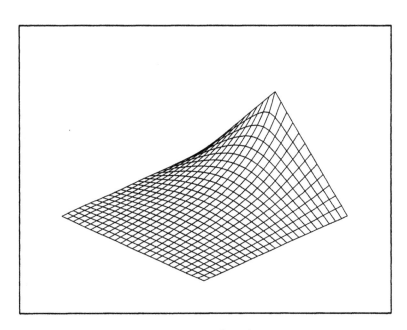

Exercise 5a. Surface $z = u(x, y)$ in perspective.

7.a. See Eq. (11). The boundary conditions require $a_n = 0$ and $c_n = \dfrac{200(1 - \cos(n\pi))}{n\pi}$.
The solution is

$$u(x, y) = \sum_{n=1}^{\infty} c_n \frac{\sinh(\lambda_n y)}{\sinh(\lambda_n b)} \sin(\lambda_n x)$$

b. This problem has to be broken into two parts. Part 1: $u_1(x, y)$ satisfies the problem
stated in a. above. Part 2: $u_2(x, y)$ satisfies the potential equation in the rectangle,

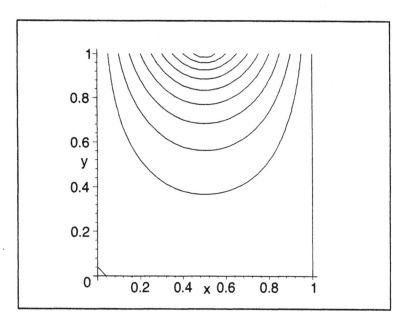

Exercise 5b. Surface $z = u(x, y)$ as a contour plot.

and $u_2(a, y) = 100$, $u_2 = 0$ on the other three sides. The solution can be obtained by exchanging x and y, a and b in the previous solution. Then

$$u_2(x, y) = \sum_{n=1}^{\infty} c_n \frac{\sinh(\mu_n x)}{\sinh(\mu_n a)} \sin(\mu_n y)$$

where $\mu_n = n\pi/b$, and

$$c_n = \frac{200(1 - \cos(n\pi))}{n\pi}.$$

Then $u(x, y) = u_1(x, y) + u_2(x, y)$.

c. This problem also must be split into two parts. Both $u_1(x, y)$ and $u_2(x, y)$ satisfy the potential equation in the rectangle. The boundary conditions are:

$$u_1(0, y) = 0, \quad u_1(a, y) = 0, \quad 0 < y < b$$

$$u_1(x, 0) = 0, \quad u_1(x, b) = bx, \quad 0 < x < a$$

and

$$u_2(0, y) = 0, \quad u_2(a, y) = ay, \quad 0 < y < b$$

$$u_2(x, 0) = 0, \quad u_2(x, b) = 0, \quad 0 < x < a$$

The problem for u_1 is solved in the text, Eq. (11). The coefficients are $a_n = 0$, $c_n = 2ab(-1)^{n+1}/n\pi$. The solution to the problem for u_2 is obtained by exchanging x and y, a and b. (Notice that this exchange also changes the boundary conditions correctly!). Finally $u(x, y) = u_1(x, y) + u_2(x, y)$,

$$u_1(x, y) = \sum_{n=1}^{\infty} c_n \frac{\sinh(\lambda_n y)}{\sinh(\lambda_n b)} \sin(\lambda_n x)$$

$$u_2(x, y) = \sum_{n=1}^{\infty} c_n \frac{\sinh(\mu_n x)}{\sinh(\mu_n a)} \sin(\mu_n y)$$

$\lambda_n = n\pi/a$, $\mu_n = n\pi/b$.

Note that $u(x, y) = xy$ is a much easier way to express the solution.

Chapter 4

4.3 Further Examples for a Rectangle

1.a. See Answers. $u(x, y) \equiv 1$ obviously satisfies all conditions.

b. The separation of variables leads to the eigenvalue problem $X'' + \lambda^2 X = 0$, $X'(0) = 0$, $X'(a) = 0$, because there are homogeneous conditions on the facing left and right sides. The solution is $X_0(x) = 1$, $\lambda_0 = 0$; $X_n(x) = \cos(\lambda_n x)$, $\lambda_n = \dfrac{n\pi}{a} (n = 1, 2, \cdots)$.

Next, solve $Y'' - \lambda^2 Y = 0$ for the values of λ found. For $\lambda = \lambda_n = \dfrac{n\pi}{a}$, $Y_n(y) = a_n \cosh(\lambda_n y) + b_n \sinh(\lambda_n y)$, $n = 1, 2, \cdots$. However for $\lambda = \lambda_0$, the differential equationfor Y is just $Y'' = 0$, with solution $Y_0(y) = a_0 + b_0 y$. The series solution is thus

$$u(x, y) = a_0 + b_0 y + \sum_1^\infty (a_n \cosh(\lambda_n y) + b_n \sinh(\lambda_n y)) \cos(\lambda_n x).$$

Now apply the conditions at $y = 0$ and $y = b$.

$$u(x, 0) = 0: \quad a_0 + \sum_1^\infty a_n \cos(\lambda_n x) = 0, \quad 0 < x < a$$

Hence $a_0 = 0$, $a_n = 0$.

$$u(x, b) = 1: \quad b_0 b + \sum_1^\infty b_n \sinh(\lambda_n b) \cos(\lambda_n x) = 1, \quad 0 < x < a.$$

This is a Fourier cosine series, whose coefficients are

$$b_0 b = \frac{1}{a} \int_0^a 1 \cdot dx = 1$$

$$b_n \sinh(\lambda_n b) = \frac{2}{a} \int_) ^a 1 \cdot \cos(\lambda_n x) dx = 0.$$

Thus, the solution boils down to $u(x, y) = \dfrac{y}{b}$.

c. In this case, the facing sides with homogeneous conditions are at $y = 0, b$. The eigenvalue problem is $Y'' + \lambda^2 Y = 0$, $Y'(0) = 0$, $Y(b) = 0$, with solution $Y_n(y) = \cos(\lambda_n y)$, $\lambda_n = \left(n - \dfrac{1}{2}\right)\dfrac{\pi}{b}$, $n = 1, 2, \cdots$. The other functions are

$$X_n(x) = a_n \cosh(\lambda_n x) + b_n \sinh(\lambda_n x).$$

The solution we seek is

$$u(x, y) = \sum_{n=1}^\infty (a_n \cosh(\lambda_n x) + b_n \sinh(\lambda_n x)) \cos(\lambda_n y).$$

Now apply the remaining boundary conditions:

$$u(0, y) = 1 : \sum_{n=1}^{\infty} a_n \cos(\lambda_n y) = 1, \quad 0 < y < b.$$

Hence

$$a_n = \frac{2}{b} \int_0^b \cos(\lambda_n y) dy = \frac{2 \sin\left((n - \frac{1}{2})\pi\right)}{(n - \frac{1}{2})\pi}$$

$$u(a, y) = 0 : \sum_{n=1}^{\infty} X_n(a) \cos(\lambda_n y) = 0, \quad 0 < y < b.$$

Hence $X_n(a) = 0$ or

$$a_n \cosh(\lambda_n a) + b_n \sinh(\lambda_n a) = 0,$$

or

$$b_n = -a_n \frac{\cosh(\lambda_n a)}{\sinh(\lambda_n a)}.$$

Now,

$$X_n(x) = a_n \left(\cosh(\lambda_n x) - \frac{\cosh(\lambda_n a)}{\sinh(\lambda_n a)} \sinh(\lambda_n x) \right) = a_n \frac{\sinh(\lambda_n(a - x))}{\sinh(\lambda_n a)}.$$

See hyperbolic identities in the Mathematical References.

3. $b_0 b = V_0/2$, $b_n \sinh(\lambda_n b) = 2V_0(\cos(n\pi) - 1)/n^2\pi^2$

$$u(x, y) = \frac{V_0}{2} \frac{y}{b} + \sum_{n=1}^{\infty} \frac{2V_0(\cos(n\pi) - 1)}{n^2\pi^2} \frac{\sinh(\lambda_n y)}{\sinh(\lambda_n b)} \cos(\lambda_n x)$$

5. The given product solution is

$$u_n = \cos(\mu_n y)(A_n \cosh(\mu_n x) + B_n \sinh(\mu_n x)).$$

The derivative $\partial u_n/\partial y$ contains a $\sin(\mu_n y)$ as a factor, hence is 0 at $y = 0$. Also, at $y = b$, u_n is 0 because of the cosine factor: $\cos(\mu_n b) = \cos((n - \frac{1}{2})\pi) = 0$.

At $x = 0$, we need $\partial u/\partial x = 0$

$$\frac{\partial u_n}{\partial x} = \cos(\mu_n y) \cdot (A_n \mu_n \sinh(\mu_n x) + B_n \mu_n \cosh(\mu_n x)).$$

At $x = 0$, this becomes $B_n \mu_n \cos(\mu_n y)$. Therefore $B_n = 0$. Now, the series for u_2 is

$$u_2(x, y) = \sum_{n=1}^{\infty} A_n \cosh(\mu_n x) \cos(\mu_n y).$$

At $x = a$, this is to be Sy, so

$$\sum_{n=1}^{\infty} A_n \cosh(\mu_n a) \cos(\mu_n y) = Sy, \quad 0 < y < b$$

$$A_n \cosh(\mu_n a) = \frac{2}{b} \int_0^b Sy \cos(\mu_n y) dy = 2Sb \left[\frac{1}{(n - \frac{1}{2})^2 \pi^2} + \frac{\sin((n - \frac{1}{2})\pi)}{(n - \frac{1}{2})\pi} \right] = c_n.$$

Finally,

$$u_2(x, y) = \sum_{n=1}^{\infty} c_n \frac{\cosh(\mu_n x)}{\cosh(\mu_n a)} \cos(\mu_n y).$$

7. The problem has homogeneous boundary conditions at $x = 0$ and $x = a$. The eigenvalue problem is

$$X'' + \lambda^2 X = 0, \quad 0 < x < a$$

$$X'(0) = 0, \quad X(a) = 0$$

with $X_n(x) = \cos(\lambda_n x)$, $\lambda_n = (n - \frac{1}{2})\pi/a$. Also $Y_n'' - \lambda_n^2 Y_n = 0$ and $Y_n'(0) = 0$, so $Y_n(y) = \cosh(\lambda_n y)$. Then

$$w(x, y) = \sum_{n=1}^{\infty} a_n \cosh(\lambda_n y) \cos(\lambda_n x).$$

The remaining condition, at $y = b$, is $w(x, b) = Sb(x - a)/a$, or

$$\sum_{n=1}^{\infty} a_n \cosh(\lambda_n b) \cos(\lambda_n x) = Sb(x - a)/a$$

$$a_n \cosh(\lambda_n b) = \frac{2}{a} \int_0^a \frac{Sb}{a}(x - a) \cos(\lambda_n x) dx = -\frac{2Sb}{(n - \frac{1}{2})^2 \pi^2}.$$

9. This is a routine problem from Section 4.2. The solution has the form

$$w(x, y) = \sum_{n=1}^{\infty} \frac{c_n \sinh(\lambda_n y) + a_n \sinh(\lambda_n (b - y))}{\sinh(\lambda_n b)} \sin(\lambda_n x).$$

The coefficients are

$$a_n = c_n = \frac{2}{a} \int_0^a -\frac{Hx(a - x)}{2} \sin(\lambda_n x) dx = -2Ha^2 \frac{1 - \cos(n\pi)}{n^3 \pi^3}.$$

11. Calculate directly

$$\nabla^2 p = (12Ax^2 + 6Bxy + 2Cy^2) + (2Cx^2 + 6Dxy + 12Ey^2)$$

and equate to $-H = -K(x^2 + y^2)$. Then $12A + 2C = -K$, $12E + 2C = -K$. These are two equations in three unknowns A, C, E. Many solutions are possible, especially

$$C = -K/2, A = E = 0 : \quad p = -Kx^2y^2/2$$

and

$$C = 0, A = E = -K/12 : \quad p = -K(x^4 + y^4)/12$$

Chapter 4

4.4 Potential in Unbounded Regions

1. The only remaining condition is the one at $y = 0$: $u_1(x, 0) = f(x)$, $\quad 0 < x < a$. With the form of Equation (7), this becomes

$$\sum_1^\infty a_n \sin\left(\frac{n\pi x}{a}\right) = f(x), \quad 0 < x < a.$$

Hence $a_n = \dfrac{2}{a} \displaystyle\int_0^a f(x) \sin\left(\frac{n\pi x}{a}\right) dx.$

3. There are conditions on u_2 at $x = 0$ and $x = a$.

$$u_2(0, y) = g_1(y) : \quad \int_0^\infty B(\mu) \sin(\mu y) d\mu = g_1(y), \quad 0 < y$$

This is a Fourier sine integral, so

$$B(\mu) = \frac{2}{\pi} \int_0^\infty g_1(y) \sin(\mu y) dy.$$

Similarly, $A(\mu) = \dfrac{2}{\pi} \displaystyle\int_0^\infty g_2(y) \cos(\mu y) dy.$

5.a. There are homogeneous conditions at $x = 0$, a, so it is not necessary to split the problem into two. In the separation of variables the ratio X''/X has to be negative. The eigenvalue problem is $X'' + \lambda^2 X = 0$, $X'(0) = 0$, $X(a) = 0$, with solution $X_n(x) = \cos(\lambda_n x)$, $\lambda_n = (n - \frac{1}{2})\pi/a$, $n = 1, 2, \cdots$. Then $Y_n'' - \lambda_n^2 Y_n = 0$ plus the boundedness as $y \to \infty$ gives $Y_n(y) = \exp(-\lambda_n y)$. The solution has the form

$$u(x, y) = \sum_{n=1}^\infty c_n \cos(\lambda_n x) \exp(-\lambda_n y).$$

The boundary condition at $y = 0$ gives $\displaystyle\sum_{n=1}^\infty c_n \cos(\lambda_n x) = 1, \, 0 < x < a.$

$$c_n = \frac{2}{a} \int_0^a 1 \cdot \cos(\lambda_n x) dx = \frac{2 \sin(\lambda_n a)}{\lambda_n a}.$$

b. In this problem, the homogeneous condition at $y = 0$ leads to the singular eigenvalue problem $Y'' + \lambda^2 Y = 0$, $Y(0) = 0$, Y bounded; solution: $Y(y; \lambda) = \sin(\lambda y)$, $\lambda > 0$. The other equation is $X'' - \lambda^2 X = 0$. We can apply the homogeneous condition $X'(0) = 0$ at this stage to determine $X(x; \lambda) = \cosh(\lambda x)$.

The solution of the potential equation that satisfies the homogeneous boundary conditions is as shown in Answers. The boundary condition at $x = a$ is $u(x, b) = e^{-y}$, $0 < y$, which becomes

$$\int_0^\infty B(\lambda) \cosh(\lambda b) \sin(\lambda y) d\lambda = e^{-y}, \quad 0 < y$$

Whatever multiplies the $\sin(\lambda y)$ in the integrand has to be the Fourier sine integral coefficient function for e^{-y}:

$$B(\lambda)\cosh(\lambda b) = \frac{2}{\pi}\int_0^\infty e^{-y}\sin(\lambda y)dy.$$

c. Again, we have a homogeneous condition at $y = 0$, leading to the eigenvalue problem $Y'' + \lambda^2 Y = 0$, $Y'(0) = 0$, Y bounded. Solution: $Y(y;\lambda) = \cos(\lambda y)$, $0 < \lambda$. For the other direction, $X'' - \lambda^2 X = 0$. If we apply the homogeneous condition $X(0) = 0$, we get $X(x;\lambda) = \sinh(\lambda x)$.

The assembled solution is

$$u(x,t) = \int_0^\infty A(\lambda)\sinh(\lambda x)\cos(\lambda y)d\lambda.$$

The nonhomogeneous boundary condition at $x = a$ becomes the Fourier cosine integral problem

$$\int_0^\infty A(\lambda)\sinh(\lambda a)\cos(\lambda y) = f(y), \quad 0 < y.$$

Hence

$$A(\lambda)\sinh(\lambda) = \frac{2}{\pi}\int_0^\infty f(y)\cos(\lambda y)dy = \frac{2}{\pi}\int_0^b \cos(\lambda y)dy = \frac{2\sin(\lambda b)}{\pi\lambda}.$$

7. Since the conditions on all three boundaries are nonhomogeneous, we need to split up the problem into two, as in the text. We would then have $f(x) = 1$, $0 < x < a$, and $g_1(y) = g_2(y) = e^{-y}$, $0 < y$. The solutions for u_1 and u_2 are as in Equations (7), (9) and Exercises 1 and 3. Details are in Answers.

9. The slot is horizontal now. In each problem, homogeneous boundary conditions are placed so that neither problem has to be split.

a. EVP: $X'' + \lambda^2 X = 0$, $X(0) = 0$, $X(x)$ bounded: $X(x;\lambda) = \sin(\lambda x)$. Then $Y'' - \lambda^2 Y = 0$, $Y(0) = 0$: $Y(y;\lambda) = \sinh(\lambda y)$. Solution:

$$u(x,y) = \int_0^\infty B(\lambda)\sinh(\lambda y)\sin(\lambda x)d\lambda.$$

Nonhomogeneous boundary conditions $u(x,b) = f(x)$, $0 < x$ becomes the Fourier sine integral problem

$$\int_0^\infty B(\lambda)\sinh(\lambda b)\sin(\lambda x)d\lambda = f(x), \quad 0 < x.$$

$$B(\lambda)\sinh(\lambda b) = \frac{2}{\pi}\int_0^\infty f(x)\sin(\lambda x)dx.$$

See Answers now.

b. The EVP is as in part a. The Y-problem is $Y'' - \lambda^2 Y = 0$, $Y(b) = 0$, with solution $Y = \sinh(\lambda(b - y))$ or its equivalent. The solution of the partial differential equation is

$$u(x, y) = \int_0^\infty B(\lambda) \sinh(\lambda(b - y)) \sin(\lambda x) d\lambda.$$

The nonhomogeneous b.c. at $y = 0$ becomes

$$\int_0^\infty B(\lambda) \sinh(\lambda b) \sin(\lambda x) d\lambda = e^{-x}, \quad 0 < x,$$

$$B(\lambda) \sinh(\lambda b) = \frac{2}{\pi} \int_0^\infty e^{-x} \sin(\lambda x) dx.$$

See Answers.

11. The region is a vertical strip. The factors in the product solution satisfy $Y'' + \lambda^2 Y = 0$, $Y(y)$ bounded as $y \to \pm\infty$; $X'' - \lambda^2 X = 0$, $X(0) = 0$. The singular eigenvalue problem for Y has both $\sin(\lambda y)$ and $\cos(\lambda y)$ as solutions; $X(x) = \sinh(\lambda x)$. Thus, the general product solution is $(A(\lambda) \cos(\lambda y) + B(\lambda) \sin(\lambda y)) \sinh(\lambda x)$, and the assembled solution is

$$u(x, y) = \int_0^\infty (A(\lambda) \cos(\lambda y) + B(\lambda) \sin(\lambda y)) \sinh(\lambda x) d\lambda.$$

Applying the boundary condition at $x = a$ gives

$$\int_0^\infty (A(\lambda) \cos(\lambda y) + B(\lambda) \sin(\lambda y)) \sinh(\lambda a) d\lambda = e^{-|y|}.$$

Since $e^{-|y|}$ is even, $B(\lambda) = 0$ and

$$A(\lambda) \sinh(\lambda a) = \frac{2}{\pi} \int_0^\infty e^{-y} \cos(\lambda y) dy.$$

13. Because there is a homogeneous condition at $x = 0$ and boundedness as $x \to \infty$, the (singular) EVP is $X'' + \lambda^2 X = 0$, $X(0) = 0$; solution $X(x; \lambda) = \sin(\lambda x)$. In the other direction, $Y'' - \lambda^2 Y = 0$, $Y(y)$ bounded, gives $Y(y; \lambda) = e^{-\lambda y}$. Combine product solutions to get

$$u(x, y) = \int_0^\infty B(\lambda) e^{-\lambda y} \sin(\lambda x) d\lambda.$$

The nonhomogeneous condition at $y = 0$ requires

$$\int_0^\infty B(\lambda) \sin(\lambda x) d\lambda = f(x), \quad 0 < x$$

hence

$$B(\lambda) = \frac{2}{\pi} \int_0^\infty f(x) \sin(\lambda x) dx.$$

15. The solution must remain bounded as $x \to \pm\infty$ and as $y \to \infty$. The (singular) EVP is $X'' + \lambda^2 X = 0$, with both $\cos(\lambda x)$ and $\sin(\lambda x)$ as solutions. In the other direction $Y'' - \lambda^2 Y = 0$ and $Y(y)$ bounded requires $Y(y) = e^{-\lambda y}$.

Combine product solutions to find

$$u(x, y) = \int_0^\infty (A(\lambda)\cos(\lambda x) + B(\lambda)\sin(\lambda x)) e^{-\lambda y}, d\lambda.$$

The boundary condition at $y = 0$ becomes

$$\int_0^\infty (A(\lambda)\cos(\lambda x) + B(\lambda)\sin(\lambda x)) d\lambda = f(x), \quad -\infty < x < \infty.$$

This is a full Fourier integral problem.

17. $u(x, y) = \dfrac{1}{\pi} \displaystyle\int_0^\infty \dfrac{y}{y^2 + (x - x')^2} dx'$. Make the substitution $u = \dfrac{(x' - x)}{y}$, $\quad du = \dfrac{dx'}{y}$. Then

$$u(x, y) = \dfrac{1}{\pi} \int_{\frac{-x}{y}}^\infty \dfrac{y}{y^2 + y^2 u^2} y\, du = \dfrac{1}{\pi} \int_{\frac{-x}{y}}^\infty \dfrac{du}{1 + u^2} = \dfrac{1}{\pi} \tan^{-1} u \Big|_{\frac{-x}{y}}^\infty$$

$$= \dfrac{1}{\pi} \left[\dfrac{\pi}{2} + \tan^{-1} \dfrac{x}{y} \right].$$

19. For the given function, $u_{xx} = u_{yy} = 0$, so it is a solution of the potential equation. By substitution, we find $u(x, 0) = x$, $u(0, y) = 0$, $u(a, y) = a$. As $x \to \infty$, $u(x, y) \to \infty$ as well, so the methods of this section do not apply – the function is not bounded.

Chapter 4

4.5 Potential in a Disk

1. The solution is given in Equation (10) with coefficients in Equation (11). The function $f(\theta) = |\theta|$, $-\pi < \theta < \pi$, is even, so $b_n = 0$. By geometry, $a_0 = \dfrac{\pi}{2}$.

$$a_n = \frac{2}{c^n \pi} \int_0^\pi \theta \cos(n\theta)d\theta = -\frac{2(1 - \cos(n\pi))}{c^n \pi n^2}.$$

3. The solution is given in Equation (10) with coefficients in Equation (11). The given function $f(\theta)$ is even, so $b_n = 0$.

$$a_n = \frac{2}{\pi c^n} \int_0^{\pi/2} \cos(\theta) \cos(n\theta)d\theta = \frac{2}{\pi c^n} \left[\frac{\sin((n-1)\theta)}{2(n-1)} + \frac{\sin((n+1)\theta)}{2(n+1)} \right]\Bigg|_0^{\pi/2}$$

$$= \frac{1}{\pi c^n} \left[\frac{\sin((n-1)\pi/2)}{n-1} + \frac{\sin((n+1)\pi/2)}{n+1} \right].$$

The numerators can be simplified using $\sin\left(\dfrac{n\pi}{2} \pm \dfrac{\pi}{2}\right) = \pm \cos\left(\dfrac{n\pi}{2}\right)$.

$$a_n = -\frac{\cos(n\pi/2)}{\pi c^n} \frac{2}{n^2 - 1}.$$

Since this formula requires division by 0 if $n = 1$, a_1 has to be calculated separately.

5. For $f(\theta)$ as specified, the series for $v(c, \theta)$ converges uniformly. See Section 1-4.

7. The boundary condition is $v(c_1\theta) = f(\theta)$ or

$$a_0 + \sum_{n=1}^\infty \frac{a_n}{c^n} \cos(n\theta) + \frac{b_n}{c^n} \sin(n\theta) = f(\theta).$$

See Answers.

9. For this case, the rays $\theta = 0$, $\theta = \dfrac{\pi}{2}$ are effectively "facing sides". Thus we have the eigenvalue problem $Q'' + \lambda^2 Q = 0$, $Q(0) = 0$, $Q(\pi/2) = 0$, with solution $Q_n(\theta) = \sin(\lambda_n \theta)$, $\lambda_n = 2n$, $n = 1, 2, \cdots$. In the r-direction, $r^2 R'' + rR' - (2n)^2 R = 0$. The general solution is a combination of r^{2n} and r^{-2n}. The latter is unbounded as $r \to 0$, so $R_n(r) = r^{2n}$. Assemble the product solutions to find

$$u(r, \theta) = \sum_{n=1}^\infty b_n r^{2n} \sin(2n\theta).$$

Now apply the boundary condition $u(c, \theta) = 1$, $0 < \theta < \frac{\pi}{2}$:

$$\sum_{n=1}^\infty b_n c^{2n} \sin(2n\theta) = 1, \quad 0 < \theta < \frac{\pi}{2}.$$

This is a routine Fourier sine series problem. We must have $b_n c^{2n}$ equal to the Fourier sine coefficient of the given function:

$$b_n c^{2n} = \frac{2}{\frac{\pi}{2}} \int_0^{\frac{\pi}{2}} 1 \cdot \sin(2n\theta) d\theta = \frac{2(1 - \cos(n\pi))}{n\pi}.$$

11. The product solution that are bounded as $r \to 0+$ are

$$v_n(r, \theta) = R_n(r)Q_n(\theta) = r^{\frac{n}{\alpha}} \sin\left(\frac{n\theta}{\alpha}\right).$$

For $n = 1$, $v_1(r, \theta) = r^{\frac{1}{\alpha}} \sin\left(\frac{\theta}{\alpha}\right)$ has $\dfrac{\partial v_1}{\partial r} = \dfrac{1}{\alpha} r^{(\frac{1}{\alpha} - 1)} \sin\left(\dfrac{\theta}{\alpha}\right)$. Since $\dfrac{1}{\alpha} - 1$ is negative, $\dfrac{\partial v_1}{\partial r}$ is unbounded as $r \to 0+$. (Test this out with $\alpha = \frac{3}{2}$.)

Chapter 4

4.6 Classification and Limitations

1.

	A	B	C	$B^2 - 4AC$	type
a	0	1	0	1	hyperbolic
b	1	1	1	-3	elliptic
c	1	-1	1	-3	elliptic
d	1	-2	1	0	parabolic
e	1	0	-1	4	hyperbolic

3. In all cases, you can find *some* product solution (b. has to be made homogeneous first). The real question is: does a suitable eigenvalue problem arise from separation of variables? The presence of the mixed partial derivative in a. through d. gets in the way.

   Take the homogeneous version of b. for instance. If $u = X(x)Y(y)$, the equation $u_{xx} + u_{xy} + u_{yy} = 0$ becomes (after dividing by XY)

   $$\frac{X''}{X} + \frac{X'\,Y'}{X\,Y} + \frac{Y''}{Y} = 0.$$

   We are no longer *forced* to conclude that, say, $\frac{Y''}{Y}$ is constant. We may still *try* to find special solutions by assuming, say, $Y'/Y = \pm\lambda$. Then $Y''/Y = \lambda^2$, $(Y' = \pm\lambda Y$, so $Y'' = \pm\lambda Y' = (\pm\lambda)^2 Y)$ and $X'' \pm \lambda X' + \lambda^2 X = 0$. This is not part of a regular Sturm-Liouville problem. Even more serious is the fact that Y satisfies a first-order differential equation and therefore will not supply enough coefficients to satisfy two conditions.

5. In each of the three, we look for solutions in the product form $u(x,y) = X(x)Y(y)$. In all three we have homogeneous boundary conditions at $x = 0, 1$. In all three we get the same EVP: $X'' + \lambda^2 X = 0$, $X(0) = 0$, $X(1) = 0$, with solution $X_n(x) = \sin(\lambda_n x)$, $\lambda_n = n\pi$, $n = 1, 2, \cdots$. The differences come from the problem for Y:

a. $Y_n'' - \lambda_n^2 Y_n = 0$, Y_n bounded as $y \to \infty$; $Y_n(y) = e^{-\lambda_n y}$.

$$u(x,y) = \sum_{n=1}^{\infty} b_n e^{-\lambda_n y} \sin(\lambda_n x)$$

b. $Y_n'' + \lambda_n^2 Y_n = 0$, $\quad Y_n(y) = b_n \cos(\lambda_n y) + a_n \sin(\lambda_n y)$.

$$u(x,y) = \sum_{n=1}^{\infty} (b_n \cos(\lambda_n y) + a_n \sin(\lambda_n y)) \sin(\lambda_n x)$$

c. $Y_n' = \lambda_n^2 Y_n, \quad Y_n(y) = e^{-\lambda_n^2 y}$

$$u(x, y) = \sum_{n=1}^{\infty} b_n e^{-\lambda_n^2 y} \sin(\lambda_n x).$$

In all three cases, applying the condition $u(x, 0) = f(x)$, $0 < x < 1$, leads to a Fourier series problem and requires

$$b_n = 2 \int_0^1 f(x) \sin(\lambda_n x) dx.$$

In problem b., the second condition at $y = 0$ makes $a_n = 0$. (Note that a_n and b_n are reversed from previous usage in the wave equation.

Comparison of solutions: Solutions of a. and c: c. decays with y more quickly, since $\lambda_n^2 > \lambda_n$. The solution of b. is a periodic function of y; it does not decay with large y.

7. Assume that $u(x, t) = X(x)T(t)$ as usual. The partial differential equation becomes

$$X''T = XT'' - \epsilon X''T''$$

or

$$\frac{X''}{X} = \frac{T^n}{T + \epsilon T''} = -\lambda^2$$

Boundary conditions give an eigenvalue problem for $X(x)$ that determines λ^2. Then the problem for $T(t)$ becomes

$$T'' + \frac{\lambda^2}{1 + \epsilon\lambda^2} T = 0.$$

Chapter 4

Miscellaneous Exercises

1. Since there are homogeneous conditions on the facing sides $y = 0$, $y = b$ the eigenvalue problem is $Y'' + \lambda^2 Y = 0$, $Y(0) = 0$, $Y(b) = 0$.

The solution is

$$u(x, y) = \sum_{n=1}^{\infty} (a_n \cosh(\lambda_n x) + b_n \sinh(\lambda_n x)) \sin(\lambda_n y)$$

with $\lambda_n = \dfrac{n\pi}{b}$, $n = 1, 2, \cdots$. Apply the b.c. at $x = 0$ to find

$$\sum_{1}^{\infty} a_n \sin(\lambda_n y) = 1, \quad 0 < y < b, \quad a_n = \frac{2(1 - \cos(n\pi))}{n\pi}.$$

Apply at $x = a$ to find

$$\sum_{n=1}^{\infty} (a_n \cosh(\lambda_n a) + b_n \sinh(\lambda_n a)) \sin(\lambda_n x) = 0, \quad 0 < x < a.$$

Thus

$$b_n = -a_n \frac{\cosh(\lambda_n a)}{\sinh(\lambda_n a)}.$$

Another way to write the solution is

$$u(x, y) = \sum_{n=1}^{\infty} a_n \frac{\sinh(\lambda_n(a - x))}{\sinh(\lambda_n a)} \sin(\lambda_n y).$$

3. The solution is $u(x, y) = 1$. This is also found by separation of variables if done correctly.

The separation of variables solution form is

$$u(x, y) = a_0 + b_0 x + \sum_{1}^{\infty} (a_n \cosh(\lambda_n x) + b_n \sinh(\lambda_n x)) \cos(\lambda_n y)$$

with $\lambda_n = n\pi/b$. The eigenvalue problem is $Y'' + \lambda^2 Y = 0$, $Y'(0) = 0$, $Y'(b) = 0$, and $\lambda_0 = 0$ is one of the eigenvalues with eigenfunction $Y_0(y) = 1$. The equation for X is $X'' - \lambda^2 X = 0$. If $\lambda = 0$, this becomes $X'' = 0$, with solutions 1 and x. Thus the product solution corresponding to $\lambda_0 = 0$ is $a_0 + b_0 x$.

Applying the boundary condition at $x = 0$ gives

$$a_0 + \sum_{1}^{\infty} a_n \cos(\lambda_n y) = 1, \quad 0 < y < b$$

from which $a_0 = 1$, $a_n = 0$. At $x = a$, we have

$$1 + b_0 \cdot b + \sum_{1}^{\infty} b_n \sinh(\lambda_n a) \cos(\lambda_n y) = 1, \quad 0 < y < b.$$

Thus $1 + b_0 b = 1$ and $b_0 = b_n = 0$.

5. Homogeneous conditions are at $y = 0$ and $y = b$. Therefore the EVP is $Y'' + \lambda^2 Y = 0$, $Y'(0) = 0$, $Y(b) = 0$; solution $Y_n(y) = \cos(\lambda_n y)$, $\lambda_n = \dfrac{(n - \frac{1}{2})\pi}{b}$. The solution has the form

$$u(x, y) = \sum_{n=1}^{\infty} (a_n \cosh(\lambda_n x) + b_n \sinh(\lambda_n x)) \cos(\lambda_n y).$$

Apply the boundary condition at $x = 0$:

$$u(x, 0) = 1 : \quad \sum_1^{\infty} a_n \cos(\lambda_n y) = 1, \quad 0 < y < b$$

Hence

$$a_n = \frac{2}{b} \int_0^b \cos(\lambda_n y) dy = \frac{2 \sin \left((n - \frac{1}{2})\pi\right)}{\left(n - \frac{1}{2}\right)\pi}.$$

The boundary condition at $x = a$ gives

$$\sum (a_n \cosh(\lambda_n a) + b_n \sinh(\lambda_n a)) \cos(\lambda_n y) = 1, \quad 0 < y < b$$

Therefore, $a_n \cosh(\lambda_n a) + b_n \sinh(\lambda_n a) = a_n$, so

$$b_n = a_n \frac{1 - \cosh(\lambda_n a)}{\sinh(\lambda_n a)}$$

7. This problem has nonhomogeneous boundary conditions on *adjacent* sides, so it has to be split into two problems. Let $w(x, y)$ be a solution of the potential equation with the boundary condition $w(x, 0) = f(x)$, $0 < x < a$ and w is 0 on the remaining edges of the square. This is a routine problem with solution

$$w(x, y) = \sum_{n=1}^{\infty} (a_n \cosh(\lambda_n y) + b_n \sinh(\lambda_n y)) \sin(\lambda_n x), \quad (\lambda_n = \frac{n\pi}{a}).$$

Another, better way to write the solution is

$$w(x, y) = \sum_{n=1}^{\infty} \frac{a_n \sinh(\lambda_n y) + b_n \sinh(\lambda_n (a - y))}{\sinh(\lambda_n a)} \sin(\lambda_n x).$$

In this form it is easier (less algebra) to do the boundary condition. Using the latter form, we find

$$w(x, 0) = \sum_{n=1}^{\infty} b_n \sin(\lambda_n x) = f(x), \quad 0 < x < a$$

$$w(x, a) = \sum_{n=1}^{\infty} a_n \sin(\lambda_n x) = 0, \quad 0 < x < a.$$

Thus

$$a_n = 0, \quad b_n = \frac{2}{a} \int_0^a f(x) \sin(\lambda_n x) dx.$$

(See Answers.)

Finally $u(x, y) = w(x, y) + w(y, x)$.

9. The "homogeneous-like" conditions are two boundedness conditions as $x \to \pm\infty$. The singular EVP is $X'' + \lambda^2 X = 0$, $X(x)$ bounded as $x \to \pm\infty$. Both $\sin(\lambda x)$ and $\cos(\lambda x)$ are eigenfunctions, and all $\lambda > 0$ are eigenvalues. The Y equation is $Y'' - \lambda^2 Y = 0$. If we impose the homogeneous condition at $y = b$, the y-factor is $Y(y; \lambda) = \sinh(\lambda(b-y))$. The form for the solution is then

$$u(x,y) = \int_0^\infty \sinh(\lambda(b-y))(A(\lambda)\cos(\lambda x) + B(\lambda)\sin(\lambda x))d\lambda.$$

The boundary condition at $y = 0$ is

$$\int_0^\infty \sinh(\lambda b)\,(A(\lambda)\cos(\lambda x) + B(\lambda)\sin(\lambda x))\,d\lambda = \begin{cases} 1, & |x| < a \\ 0, & |x| > a \end{cases}.$$

The given function is even, so $B(\lambda) = 0$, and

$$A(\lambda)\sinh(\lambda b) = \frac{2}{\pi}\int_0^a 1 \cdot \cos(\lambda x)dx, \quad \text{or} \quad A(\lambda) = \frac{2\sin(\lambda a)}{\lambda\pi\sinh(\lambda b)}.$$

11. The EVP is the same as in the solution of **9.** above. In the y-direction, we have $Y'' - \lambda^2 Y = 0$, $Y(y)$ bounded as $y \to \infty$ (because we are solving in the upper half-plane). Therefore $Y(y; \lambda) = e^{-\lambda y}$.

The solution has the form

$$y(x,y) = \int_0^\infty e^{-\lambda y}\,(A(\lambda)\cos(\lambda x) + B(\lambda)\sin(\lambda x))\,d\lambda$$

The boundary condition at $y = 0$ requires

$$\int_0^\infty (A(\lambda)\cos(\lambda x) + B(\lambda)\sin(\lambda x))\,d\lambda = e^{-\alpha|x|}.$$

The right-hand side is an even function, so $B(\lambda) = 0$ and

$$A(\lambda) = \frac{2}{\pi}\int_0^\infty e^{-\alpha x}\cos(\lambda x)dx.$$

(See Answers.)

13.

$$u(x,y) = \frac{1}{\pi}\int_{-\infty}^\infty \frac{y}{y^2 + (x - x')^2}dx' = \frac{-1}{\pi}\tan^{-1}\left(\frac{x - x'}{y}\right)\Big|_{x'=-\infty}^\infty = \frac{1}{\pi}\left[\frac{\pi}{2} - \left(-\frac{\pi}{2}\right)\right] = 1.$$

15. The solution was developed in Section 4-5, Equations (10) and (11). For the given function $u(c, \theta)$, we find

$$a_n = \frac{1}{\pi c^n}\int_0^\pi \cos(n\theta)d\theta = 0,$$

$$b_n = \frac{1}{\pi c^n}\int_0^\pi \sin(n\theta)d\theta = \frac{1 - \cos(n\pi)}{n\pi c^n}$$

and $a_0 = \frac{1}{2}$.

17. With the condition $u(c, \theta) = |\sin(\theta)|$, $b_n = 0$ because this is an even function. Then $a_0 = \dfrac{2}{\pi}$, $a_n = \dfrac{2}{\pi} \displaystyle\int_0^\pi \sin(\theta)\cos(n\theta)d\theta$. The integral table in Mathematical References can be used for $n \geq 2$ (also see Section 1-2, Exercise 11c). The computation for a_1 has to be done separately.

19. Assembling all the product solutions gives

$$u(r, \theta) = A_0 + B_0 \ln(r) + \sum_1^\infty \left[\left(A_n r^n + C_n r^{-n} \right) \cos(n\theta) + \left(B_n r^n + D_n r^{-n} \right) \sin(n\theta) \right]$$

The boundary conditions are

$$A_0 + B_0 \ln(a) + \sum \left[\left(A_n a^n + C_n a^{-n} \right) \cos(n\theta) + \left(B_n a^n + D_n a^{-n} \right) \sin(n\theta) \right] = 1.$$

This is a "routine" Fourier series problem. The coefficients are

$$A_0 + B_0 \ln(a) = 1, \quad A_n a^n + C_n a^{-n} = 0, \quad B_n a^n + D_n a^{-n} = 0.$$

At the other boundary, $r = b$, we get similar equations, except that the function on the right is 0, not 1. Hence

$$A_0 + B_0 \ln(b) = 0, \quad A_n b^n + C_n b^{-n} = 0, \quad B_n b^n + D_n b^{-n} = 0.$$

Each pair (A_0, B_0), (A_n, C_n), (B_n, D_n) satisfies a pair of equations from which each coefficient can be found. For example

$$A_0 + B_0 \ln(a) = 1$$

$$A_0 + B_0 \ln(b) = 0$$

give $A_0 = \dfrac{\ln(b)}{\ln(b/a)}$, $B_0 = \dfrac{-1}{\ln(b/a)}$. The remaining coefficients are 0, so

$$u(r, \theta) = \frac{\ln(b) - \ln(r)}{\ln(b/a)}.$$

21. There are homogeneous conditions at $\theta = 0, 2\pi$. The EVP is $Q'' + \lambda^2 Q = 0$, $Q(0) = 0$, $Q(2\pi) = 0$, with solution $Q_n(\theta) = \sin(\lambda_n \theta)$, $\lambda_n = n/2$, $n = 1, 2, \cdots$. In the r-direction, we have $r(rR_n')' - \lambda_n^2 R_n = 0$, with solutions $R_n(r) = r^{n/2}$, $r^{-n/2}$. The latter is not bounded as $r \to 0$ and must be discarded. Product solutions are

$$r^{\frac{n}{2}} \sin\left(\frac{n\theta}{2} \right), \quad n = 1, 2, \cdots.$$

Now see Answers.

23. This is a routine problem. See Answers.

25. The solution of **24** is $v(x, y) = xy/b$. Now $w(x, y) = v(x, y) - u(x, y)$ satisfies the potential equation in the rectangle and these boundary conditions

$$w(0, y) = 0, \quad w_x(a, y) = -\frac{y}{b}, \quad 0 < y < b$$

$$w(x, 0) = 0, \quad w(x, b) = 0, \quad 0 < x < a.$$

The solution is routine:

$$w(x, y) = \sum_{n=1}^{\infty} b_n \sinh(\lambda_n x) \sin(\lambda_n y)$$

with $\lambda_n = n\pi/b$. Here, the homogeneous condition at $x = 0$ has already been applied. For the condition at $x = a$ we need

$$w_x(x, y) = \sum_{n=1}^{\infty} b_n \lambda_n \cosh(\lambda_n x) \sin(\lambda_n y).$$

Then the condition reads

$$\sum_{n=1}^{\infty} b_n \lambda_n \cosh(\lambda_n a) \sin(\lambda_n y) = -\frac{y}{b}, \quad 0 < y < b$$

See Answers for b_n.

27. With $u = \phi_x$ and $v = \phi_y$, the first equation becomes $\phi_{xy} = \phi_{yx}$, true for any ϕ with continuous mixed partials. The second becomes

$$(1 - M^2)\phi_{xx} + \phi_{yy} = 0.$$

According to Section 4-6, we have $A = 1 - M^2$, $B = 0$, $C = 1$ and $B^2 - 4AC = -4(1 - M^2)$. This is: positive if $M < 1$, and the equation is elliptic; or negative if $M > 1$, and the equation is hyperbolic.

29. Solve the equation in **27.** above with boundedness as $x \to \pm\infty$, $y \to \infty$.

$$\phi(x, y) = \int_0^{\infty} (A(\alpha) \cos(\alpha x) + B(\alpha) \sin(\alpha x)) \cdot e^{-\beta y} d\alpha, \quad \beta = \alpha\sqrt{1 - M^2}.$$

Then

$$v(x, y) = \phi_y = \int_0^{\infty} -\beta e^{-\beta y} (A(\alpha) \cos(\alpha x) + B(\alpha) \sin(\alpha x)) \, dx$$

and the condition is $v(x, 0) = U_0 f'(x)$ or

$$\int_0^{\infty} -\beta (A(\alpha) \cos(\alpha x) + B(\alpha) \sin(\alpha x)) \, d\alpha = U_0 f'(x).$$

Then

$$-\beta A(\alpha) = \frac{U_0}{\pi} \int_{-\infty}^{\infty} f'(x) \cos(\alpha x) dx$$

$$-\beta B(\alpha) = \frac{U_0}{\pi} \int_{-\infty}^{\infty} f'(x) \sin(\alpha x) dx.$$

31. By Green's theorem,

$$\int_{\mathcal{R}}\int \nabla^2 u \, dA = \int_{\mathcal{C}} \frac{\partial u}{\partial n} ds$$

where $\mathcal{R}$ is a region of the x, y-plane and $\mathcal{C}$ is its boundary curve. For the problem at hand, $\nabla^2 u = 0$ in a rectangle $\mathcal{R}$, so the specified line integral must be 0 also.

33. Substitute directly.

35. $u = -\frac{1}{2}\ln(x^2 + y^2)$

$$\frac{\partial u}{\partial x} = -\frac{1}{2}\frac{2x}{x^2 + y^2}, \quad \frac{\partial u}{\partial y} = -\frac{1}{2}\cdot\frac{2y}{x^2 + y^2}$$

$$\frac{\partial^2 u}{\partial x^2} = -\frac{(x^2 + y^2) - x(2x)}{(x^2 + y^2)^2} = -\frac{y^2 - x^2}{(x^2 + y^2)^2}$$

$$\frac{\partial^2 u}{\partial y^2} = -\frac{(x^2 + y^2) - y(2y)}{(x^2 + y^2)^2} = -\frac{x^2 - y^2}{(x^2 + y^2)^2}$$

$$\mathbf{V} = -\frac{x\mathbf{i} + y\mathbf{j}}{x^2 + y^2}.$$

The magnitude of V gets large near the origin.

37.a. $\dfrac{1}{r}\dfrac{d}{dr}\left(r\dfrac{\partial u}{\partial r}\right) = -1, \quad u = \dfrac{-r^2}{4} + c_1\ln(r) + c_2$

b. $\dfrac{1}{r}\dfrac{\partial}{\partial r}\left(r\dfrac{\partial u}{\partial r}\right) = -\dfrac{1}{r^2}.$ Multiply by r and integrate; then divide by r and integrate again.

$$u(r) = -\frac{(\ln(r))^2}{2} + c_1\ln(r) + c_2$$

39. The solution has the form

$$V(x, y) = a_0 + \sum_{n=1}^{\infty} a_n e^{-\lambda_n y}\cos(\lambda_n x),$$

$\lambda_n = n\pi/L$, and coefficients determined by

$$a_0 + \sum_{1}^{\infty} a_n \cos(\lambda_n x) = f(x), \quad 0 < x < L.$$

At any $y > 5L$, $e^{-\lambda_n y} \cong 0$ for every n, so

$$V(x, y) \cong a_0 = \frac{1}{a}\int_0^a f(x)dx,$$

the average value of f.

41. Assume that $\theta(X, Y) = \phi(X)\psi(Y)$. Apply to the partial differential equation to get

$$\gamma\phi'\psi = \phi''\psi + \phi\psi''$$

or

$$\frac{\psi''}{\psi} = -\frac{\phi'' - \gamma\phi'}{\phi} = -\lambda^2$$

Boundary conditions are $\psi'(0) = 0$, $\psi'(1) = 0$, so we find $\lambda_0 = 0$, $\psi_0(Y) = 1$, $\lambda_n = n\pi$, $\psi_n(Y) = \cos(\lambda_n Y)$. The problem for $\phi(X)$ is then $\phi_n'' - \gamma\phi_n' - \lambda_n^2\phi_n = 0$, ϕ_n bounded as $X \to \infty$. The solution is $\phi_n(x) = e^{\beta_n X}$ where $\beta_n = \gamma/2 - \sqrt{(\gamma/2)^2 + \lambda_n^2}$. Note that $\beta_n < 0$. Also $\beta_0 = 0$.

The solution now has the form

$$\theta(X, Y) = a_0 + \sum_{n=1}^{\infty} a_n e^{\beta_n X} \cos(\lambda_n Y).$$

The condition at $X = 0$ is $\theta(0, Y) = 1$, from which $a_0 = 1$, $a_n = 0$. That is, $\theta(X, Y) = 1$.

Chapter 5

5.1 Two-Dimensional Wave Equation: Derivation

1. See Answers

3. See Answers

Chapter 5

5.2 Three-Dimensional Heat Equations: Vector Derivation

1. See Answers.

3. The problem for $W(x, z)$ is

$$W_{xx} + W_{zz} + \mu^2(T_2 - W) = 0, \quad 0 < x < a, \quad 0 < z < c$$

$$W(0, z) = T_0, \quad W(a, z) = T_1, \quad 0 < z < c$$

$$W_z(x, 0) = 0, \quad W_z(x, c) = 0, \quad 0 < x < a.$$

where $\mu^2 = 2h/b\kappa$. However, there is no z-variation in W: that is, $W(x, z) = W(x)$, because (1) the boundary conditions at $x = 0$ and $x = a$ have no z-variation, (2) the inhomogeneity in the differential equation has no z-variation, and (3) the boundary conditions at $z = 0$ and $z = c$ are zero-slope conditions. See Answers for the solution.

5. See Answers.

Chapter 5

5.3 Two-Dimensional Heat Equation: Solution

1. The eigenvalues are, from Equation (13) and what follows, $\lambda_{mn}^2 = \left(\frac{m\pi}{a}\right)^2 + \left(\frac{n\pi}{b}\right)^2$. If $a = b$, $\lambda_{mn}^2 = (m^2 + n^2)(\pi/a)^2$. The first four terms of the series in Equation (17) are (remember that $a = b$):

$$\frac{4a^2}{\pi^2}\left[\sin\left(\frac{\pi x}{a}\right)\sin\left(\frac{\pi y}{a}\right)\exp(-\lambda_{11}^2 kt) - \frac{1}{2}\left(\sin\left(\frac{2\pi x}{a}\right)\sin\left(\frac{\pi y}{a}\right) + \sin\left(\frac{\pi x}{a}\right)\sin\left(\frac{2\pi y}{a}\right)\right)\right.$$

$$\left.\cdot \exp(-\lambda_{21}^2 kt) + \frac{1}{4}\sin\left(\frac{2\pi x}{a}\right)\sin\left(\frac{2\pi y}{a}\right)\exp(-\lambda_{22}^2 kt)\right]$$

3. See Answers.

5. With the new boundary conditions, the EVP for X is $X'' + \mu^2 X = 0$, $X'(0) = 0$, $X'(a) = 0$. The solution: $\lambda_0 = 0$, $X_0(x) = 1$; $\mu_m = m\pi/a$, $X_m(x) = \cos(\mu_m x)$. The eigenvalues λ_{mn} are in Answers.

   The solution is as given in Equation (14), except that the index m should start at 0. You could also write

$$u(x, y, t) = \sum_{n=1}^{\infty} a_{0n}\sin(\nu_n y)\exp(-\nu_n^2 kt) + \sum_{m=1}^{\infty}\sum_{n=1}^{\infty} a_{mn}\cos(\mu_m x)\sin(\nu_n y)\exp(-\lambda_{mn}^2 kt)$$

   (Note that $\lambda_{0n}^2 = \nu_n^2$.)

7. The problem to solve is Equation (1), (4) plus boundary conditions

$$u_y(x, 0, t) = 0, \quad u_y(x, b, t) = 0, \quad 0 < x < a, \quad 0 < t$$

$$u_x(0, y, t) = 0, \quad u_x(a, y, t) = 0, \quad 0 < y < b, \quad 0 < t.$$

   The solution is best left almost at in Equation (14):

$$u(x, y, t) = \sum_{m=0}^{\infty}\sum_{n=0}^{\infty} a_{mn}\phi_{mn}(x, y)T_{mn}(t)$$

$$\phi_{0n} = \cos\left(\frac{n\pi y}{b}\right), \quad \phi_{m0} = \cos\left(\frac{m\pi x}{a}\right),$$

$$\phi_{mn} = \cos\left(\frac{m\pi x}{a}\right)\cos\left(\frac{n\pi y}{b}\right),$$

$$a_{00} = \frac{1}{ab}\int_0^a\int_0^b f(x, y)\,dy\,dx,$$

$$a_{0n} = \frac{2}{ab}\int_0^a\int_0^b f(x, y)\cos\left(\frac{n\pi y}{b}\right)dy\,dx$$

$$a_{m0} = \frac{2}{ab}\int_0^a\int_0^b f(x, y)\cos\left(\frac{m\pi x}{a}\right)dy\,dx$$

$$a_{mn} = \frac{4}{ab}\int_0^a\int_0^b f(x, y)\cos\left(\frac{m\pi x}{a}\right)\cos\left(\frac{n\pi y}{b}\right)dy\,dx.$$

The actual coefficients are given in Answers.

9. If $X''/X = +p^2$, then X is a combination of $\cosh(px)$ and $\sinh(px)$. Any such combination can cross the x-axis at most once, unless it is identically zero. Thus, the boundary conditions cannot be satisfied if $X''/X > 0$. Similar reasoning applies to $Y(y)$.

11. $u_{mn}(x, y, t) = 0$ for fixed x, y but all t means $\sin(\frac{m\pi x}{a}) = 0$ or $\sin(\frac{n\pi y}{b}) = 0$.

For $m = 1$, $n = 2$, $u_{12}(x, y, t) = 0$ for all t if and only if $x = 0$ or a, or $y = 0$, $b/2$ or b.

For $m = 2$, $n = 3$, $u_{23}(x, y, t) = 0$ for all t if and only if $x = 0$, $a/2$, or a, or $y = 0$, $b/3$, $2b/3$, or b.

In general, the nodal lines for $u_{mn}(x, y, t) = 0$ form a grid of m vertical and n horizontal lines (including the bounding lines).

Chapter 5

5.4 Problems in Polar Coordinates

1. See Answers.

3. $v(r, \theta, t) = \phi(r, \theta)T(t)$. Since $\nabla^2 v = T(t)\nabla^2\phi$ and $v_t = \phi T'$, $v_{tt} = \phi T''$, the factor $T(t)$ satisfies $T' = -\lambda^2 kT$ for the heat equation or $T'' = -\lambda^2 c^2 T$ for the wave equation.

5. Equation (9) is unchanged. The conditions of Equation (10), (11) are replaced by homogeneous boundary conditions at $\theta = 0, \pi$. The EVP is $\Theta'' + \lambda^2\Theta = 0$, $\Theta(0) = 0$, $\Theta(\pi) = 0$. See Answers for the solution.

7. Let ϕ_m, ϕ_k be eigenfunctions belonging to different eigenvalues.

$$\nabla^2\phi_k = -\lambda_k^2\phi_k$$

$$\nabla^2\phi_m = -\lambda_m^2\phi_m$$

The integrand on the left is

$$\phi_k\nabla^2\phi_m - \phi_m\nabla^2\phi_k = (\lambda_k^2 - \lambda_m^2)\phi_m\phi_k.$$

The equality is

$$(\lambda_k^2 - \lambda_m^2)\int\int_{\mathcal{R}}\phi_m\phi_k dA = \int_{\mathcal{C}}\left(\phi_k\frac{\partial\phi_m}{\partial n} - \phi_m\frac{\partial\phi_k}{\partial n}\right)ds.$$

Since both ϕ_k and ϕ_m are 0 on $\mathcal{C}$, the right-hand side is 0. Also $\lambda_k^2 - \lambda_m^2 \neq 0$. Consequently

$$\int\int_{\mathcal{R}}\phi_m\phi_k dA = 0$$

if $\lambda_m^2 \neq \lambda_k^2$.

Chapter 5

5.5 Bessel's Equation

1. The general solution of the differential equation is $\phi(r) = AJ_0(\lambda r) + BY_0(\lambda r)$. Since $|Y_0(\lambda r)| \to \infty$ as $r \to 0+$, the boundedness condition at $r = 0$ requires $B = 0$. The second condition is now $\phi(a) = 0$: $AJ_0(\lambda a) = 0$. The solution of the differential equationcan be nonzero only if $J_0(\lambda a) = 0$, or $\lambda a = \alpha_{01}, \alpha_{02}, \cdots$. Therefore $\lambda_n = \alpha_{0n}/a$, $\phi_n(r) = J_0(\lambda_n r)$, $n = 1, 2, \cdots$.

3. This is just the chain rule:

$$\frac{d}{dr}J_\mu(\lambda r) = \frac{d}{d(\lambda r)}J_\mu(\lambda r) \cdot \frac{d(\lambda r)}{dr}$$

$$= J'_\mu(\lambda r) \cdot \lambda$$

5. By definition of α_{0n}, $J_0(r) = 0$ if r is one of the numbers $\alpha_{0,1}, \alpha_{0,2}, \cdots$. For any n, $J_0(r) = 0$ at the points $r = \alpha_{0,n}$ and $r = \alpha_{0,n+1}$, and $J_0(r)$ and its derivative $-J_1(r)$ are continuous for all r. By Rolle's theorem, $J_1(r) = 0$ for some value of r between $\alpha_{0,n}$ and $\alpha_{0,n+1}$.

7. In the second formula of Exercise 6, replace μ with $\mu + 1$ and $\mu - 1$ with μ (raise both subscripts by 1):

$$\frac{d}{dx}\left(x^{\mu+1}J_{\mu+1}(x)\right) = x^{\mu+1}J_\mu(x).$$

Now integrate both sides. Integration "undoes" the differentiation

$$x^{\mu+1}J_{\mu+1}(x) = \int x^\mu J_\mu(x) \cdot xdx.$$

Notice the x before the dx.

9. Add the condition that $u(0)$ be bounded. Then $u_p = T$ is a particular solution (obvious by physical reasoning). The solution of the homogeneous equation that is bounded at $r = 0$ is $u_c = c_1 I_0(\gamma r)$. Thus the general solution of the nonhomogeneous differential equation subject to the boundedness at $r = 0$ is $u(r) = T + c_1 I_0(\gamma r)$. Apply the boundary condition at $r = a$ to find c_1:

$$T + c_1 I_0(\gamma a) = T_1, \quad c_1 = \frac{(T_1 - T)}{I_0(\gamma a)}.$$

Finally

$$u(r) = T + (T_1 - T)\frac{I_0(\gamma r)}{I_0(\gamma a)}.$$

Chapter 5

5.6 Temperature in a Cylinder

1. See Answers.

3. The form of the solution is given in Equation (13) and the formula for the coefficients is in Equation (12). From Equation (16), the denominator is $a^2 J_1^2(\alpha_n)/2$. The numerator is

$$\int_0^a f(r) J_0(\lambda_n r) r\, dr = \int_0^{\frac{a}{2}} T_0 J_0(\lambda_n r) r\, dr = \frac{T_0 r J_1(\lambda_n r)}{\lambda_n}\Bigg|_0^{\frac{a}{2}}$$

$$= T_0 \frac{a J_1(\lambda_n a/2)}{2\lambda_n} = T_0 \frac{a^2 J_1(\alpha_n/2)}{2\alpha_n},$$

using the fact that $\lambda_n a = \alpha_n$. Finally

$$a_n = T_0 \frac{J_1(\alpha_n/2)}{\alpha_n J_1^2(\alpha_n)}.$$

5. Take the equation of Exercise **4** one term at a time: First term

$$\int_0^a \frac{d}{dr}\left[(r\phi')^2\right] dr = (r\phi')^2\big|_0^a = [a\phi'(a)]^2$$

Second term (use integration by parts)

$$\lambda^2 \int_0^a r^2 \frac{d}{dr}[\phi^2]dr = \lambda^2\left[r^2\phi^2\big|_0^a - \int_0^a 2r\phi^2 dr\right] = -2\lambda^2 \int_0^a \phi^2 r\, dr.$$

Note that $r^2\phi^2(r) = 0$ at $r = 0$ (obvious) and at $r = a$ (because $\phi(a) = 0$). Now, the integrated equation from **4** is

$$[a\phi'(a)]^2 - 2\lambda^2 \int_0^a \phi^2 r\, dr = 0$$

and the result follows by simple algebra.

Chapter 5

5.7 Vibrations of a Circular Membrane

1. A typical function in the series is $J_0(\lambda_n r)\cos(\lambda_n ct)$, which we call $w(r,t)$ for brevity. Clearly $w_{tt} = -\lambda_n^2 c^2 w$. Also,

$$\frac{1}{r}\frac{d}{dr}\left(r\frac{d}{dr}J_0(\lambda r)\right) = -\lambda^2 J_0(\lambda r)$$

from the Bessel equation. (See the summary at the end of Section 5-5.) Therefore

$$\frac{1}{r}\frac{\partial}{\partial r}\left(r\frac{\partial w}{\partial r}\right) = -\lambda_n^2 w.$$

Thus, the partial differential equation (1) is satisfied.

Equation (8) is satisfied because $J_0(\lambda r)$ is bounded at $r = 0$. Also, because of the way λ_n was chosen in Equation (9), $w(a,t) = J_0(\alpha_n)\cos(\lambda_n ct) = 0$.

3. The multiplier of t in the sine and cosine in Equation (25) is $\lambda_{mn}c$, so the frequencies of vibration are $\dfrac{\lambda_{mn}c}{2\pi}$. The five lowest would be those corresponding to the five smallest values of $\dfrac{\alpha_{mn}}{a}$. See Answers.

5. Each function ϕ of Equation (18) must satisfy $\phi(a,\theta) = 0$ [from Equation (12)]; $\phi(0,\theta)$ bounded [13]; $\phi(r,\theta) = \phi(r,\theta + 2\pi)$ [14].

7. Follow the hint in Answers. Let $J_m(\lambda_{mn}r) = \phi_n$. The differential equation satisfied by ϕ_n is Bessel's equation: $(r\phi_n')' + \lambda_{mn}^2 r\phi_n$.

Now follow the proof of orthogonality from Section 2-7.

$$\phi_q(r\phi_n')' = -\lambda_{mn}^2 r\phi_q\phi_n$$

$$\phi_n(r\phi_q')' = -\lambda_{mq}^2 r\phi_n\phi_q.$$

Subtract and integrate 0 to a

$$\int_0^a \left[\phi_q(r\phi_n')' - \phi_n(r\phi_q')'\right] dr = \left(\lambda_{mq}^2 - \lambda_{mn}^2\right)\int_0^a \phi_q\phi_n r\,dr.$$

Use integration by parts on each term of the integral on the left; it becomes

$$\phi_q r\phi_n'\Big|_0^a - \int_0^a r\phi_n'\phi_q'dr - \phi_n r\phi_q'\Big|_0^a + \int_0^a r\phi_q'\phi_n'dr.$$

The integrals are identical except for sign. The evaluations are 0 because $\phi_n(a) = \phi_q(a) = 0$. Thus, since $\lambda_{mq}^2 \neq \lambda_{mn}^2$, we have $\int_0^a \phi_q\phi_n r\,dr = 0$.

9. The nodal curves occur where $J_0(\lambda_{0n}r) = 0$; that is, where $\lambda_{0n}r = \lambda_{0m}$ for $m \leq n$. Therefore the radii are $r = \lambda_{0m}/\lambda_{0n} = \alpha_{0m}/\alpha_{0n}$. For $n = 2$, $r = 2.405/5.520 = 0.436$, and $r = 1$. For $n = 2$, $r = 2.405/8.654 = 0.278$, $5.520/8.654 = 0.638$ and $r = 1$.

Chapter 5

5.8 Some Applications of Bessel Functions

1. Differentiate the product and divide by x^n to get

$$\phi'' + \frac{n}{x}\phi' + \lambda^2\phi = 0 \tag{$*$}$$

Now compare to Equation (1) and equate corresponding terms Equation (1) Equation ($*$)

Eq.(1)		Eq.($*$)
$1 - 2\alpha$	$=$	n
$p^2\gamma^2 - \alpha^2$	$=$	0
$\gamma - 1$	$=$	0
λ	$=$	λ

From these, $\alpha = (1-n)/2$, $\gamma = 1$, $p = |\alpha|$. Now the Answer follows.

3. Equation (9) is $Z'' - \lambda^2 Z = 0$. If $\lambda = 0$, $Z = c_1 + c_2 z$. If $\lambda > 0$, $Z = c_1 \cosh(\lambda z) + c_2 \sinh(\lambda z)$. This can be modified into a combination of $\sinh(\lambda z)$ and $\sinh(\lambda(b-z))$, as in Equation (12). See Mathematical References.

5. Follow the development of Section 3-3. The initial conditions are

$$\phi(\rho) + \psi(\rho) = \rho f(\rho) = F(\rho) \tag{15'}$$

$$c\phi'(\rho) - c\psi'(\rho) = \rho g(\rho). \tag{16'}$$

Divide through Equation (16') by c and integrate.

$$\phi(\rho) - \psi(\rho) = \int_0^\rho \frac{1}{c}xg(x)dx = G(\rho) \tag{16''}$$

The names F and G are assigned for convenience. Now solve Equation (15') and (16'') simultaneously to get $\phi(\rho) = \frac{1}{2}(F(\rho) + G(\rho))$, $\psi(\rho) = \frac{1}{2}(F(\rho) - G(\rho))$. These solutions are valid only for $0 < \rho < a$; we need extensions $\tilde{F}$ and $\tilde{G}$ that will be valid for all arguments. The extensions are determined by the remaining conditions: at $\rho = 0$, u must be bounded, so $\phi(ct) + \psi(-ct) = 0$, or

$$\tilde{F}(ct) + \tilde{G}(ct) + \tilde{F}(-ct) - \tilde{G}(-ct) = 0 \tag{$*$}$$

From the condition $u(a, t) = 0$, we have

$$\tilde{F}(a + ct) + \tilde{G}(a + ct) + \tilde{F}(a - ct) - \tilde{G}(a - ct) = 0 \tag{$**$}$$

As in Section 3-3, $\tilde{F} = \bar{F}_o$ and $\tilde{G} = \bar{G}_e$.

7. See Sections 2-7 and 5-7 for theory. The orthogonality condition is

$$\int_0^a R_m(\rho)R_n(\rho) \cdot \rho^2 d\rho = 0 \quad (m \neq n)$$

Because of Equation (21), this obviously reduces to the familiar conditions

$$\int_0^a \sin\left(\frac{n\pi\rho}{a}\right)\sin\left(\frac{m\pi\rho}{a}\right) d\rho = 0 \quad (n \neq m)$$

Applying the formula in Equation (22) to the initial conditions, we find $u(\rho,0) = f(\rho)$ becomes $\sum_1^\infty a_n \sin(\lambda_n\rho) = \rho f(\rho)$, $0 < \rho < a$ (a routine Fourier sine series for $\rho f(\rho)$,

$$a_n = \frac{2}{a}\int_0^a \rho f(\rho)\sin(\lambda_n\rho)d\rho$$

$$b_n = \frac{2}{\lambda_n ac}\int_0^a \rho g(\rho)\sin(\lambda_n\rho)d\rho.$$

9. Use integration and algebra to solve:

$$x^3 v' = -x + c_1; \quad v' = -\frac{1}{x^2} + \frac{c_1}{x^3}, \quad v = \frac{1}{x} - \frac{c_1}{2x^2} + c_2.$$

Now apply the boundary conditions

$$v(a) = 0: \quad \frac{1}{a} - \frac{c_1}{2a^2} + c_2 = 0$$

$$v(b) = 0: \quad \frac{1}{b} - \frac{c_1}{2b^2} + c_2 = 0$$

and solve for $c_1 = \dfrac{2ab}{a+b}$, $c_2 = \dfrac{-1}{a+b}$. Algebra produces the Answer.

11. The technique consists of trying to find a solution of the partial differential equation that depends on just one of the variables, x or y. If the inhomogeneity contains both variables, the technique won't work.

13. Use the orthogonality principle on the boundary condition $u(x,c) = -v(x)$, which takes the form

$$\sum_{n=1}^\infty a_n X_n(x) = -v(x), \quad a < x < b$$

From here,

$$\sum_{n=1}^\infty a_n \int_a^b X_n(x)X_m(x)x^3 dx = -\int_a^b v(x)X_m(x)x^3 dx.$$

Then, the only nonzero term of the series is the one with $n = m$, so

$$a_m \int_a^b X_m^2(x)x^3 dx = -\int_a^b v(x)X_m(x)x^3 dx.$$

This gives the equation in Answers.

Chapter 5

5.9 Spherical Coordinates; Legendre Polynomials

1. The solution is a *tour-de-force* of indexing. First, take the derivative of Φ as given and multiply by $\sin\phi$:

$$\sin(\phi)\Phi' = \sum_{k=1}^{\infty} -ka_k \sin(k\phi)\sin(\phi)$$

Next, use the identity given in the text:

$$\sin(\phi)\Phi' = \sum_{k=1}^{\infty} \frac{-ka_k}{2}\left[\cos((k-1)\phi) - \cos((k+1)\phi)\right].$$

Differentiate again:

$$(\sin(\phi)\Phi')' = \sum_{k=1}^{\infty} \frac{ka_k}{2}\left[(k-1)\sin((k-1)\phi) - (k+1)\sin((k+1)\phi)\right].$$

Now compute the second term of the equation, using the identity

$$\sin(\phi)\cos(k\phi) = \frac{1}{2}\left[\sin((k+1)\phi) - \sin((k-1)\phi)\right]$$

$$\mu^2 \sin(\phi)\Phi = \frac{\mu^2 a_0}{2}\sin\phi + \sum_{k=1}^{\infty} \frac{\mu^2 a_k}{2}\left[\sin((k+1)\phi) - \sin((k-1)\phi)\right].$$

The two terms of the differential equation are now Fourier sine series, but the indices have to be shifted – somewhat like a change of variables in definite integrals. For instance,

$$\sum_{k=1}^{\infty} \frac{ka_k}{2}(k+1)\sin((k+1)\phi) = \sum_{m=2}^{\infty} \frac{(m-1)a_{m-1}}{2}m\sin(m\phi)$$

using the change $m = k+1$. Note that the summation starts at $m = 2$.

Similarly

$$\sum_{k=1}^{\infty} \frac{ka_k}{2}(k-1)\sin((k-1)\phi) = \sum_{m=1}^{\infty} \frac{(m+1)a_{m+1}}{2}m\sin(m\phi)$$

using $m = k-1$ this time. Note that the term corresponding to $k = 1$ or $m = 2$ is actually 0, so the m-sum starts at 1. Do similar transformations on the series for the second term of the differential equation. Then the differential equation requires

$$0 = \sum_{m=1}^{\infty} \frac{(m+1)ma_{m+1}}{2}\sin(m\phi) - \sum_{m=2}^{\infty} \frac{(m-1)ma_{m-1}}{2}\sin(m\phi)$$

$$+ \frac{\mu^2 a_0}{2}\sin(\phi) + \sum_{m=2}^{\infty} \frac{\mu^2 a_{m-1}}{2}\sin(m\phi) - \sum_{m=1}^{\infty} \frac{\mu^2 a_{m+1}}{2}\sin(m\phi).$$

The net coefficient of each sine must be 0. First, collecting the coefficients of $\sin(\phi)$ we find

$$\frac{2\cdot 1 \cdot a_2}{2} + \frac{\mu^2 a_0}{2} - \frac{\mu^2 a_2}{2} = 0$$

or $(2 - \mu^2)a_2 = \mu^2 a_0$.

Next, collecting the coefficients for general $m \geq 2$ (all the series have such terms), we find

$$(m + 1)ma_{m+1} - (m - 1)ma_{m-1} + \mu^2 a_{m-1} - \mu^2 a_{m+1} = 0$$

or

$$\left[(m + 1)m - \mu^2\right] a_{m+1} = \left[(m - 1)m - \mu^2\right] a_{m-1}.$$

Observe: (1) a_0 and a_1 are arbitrary, as you might expect for a solution of a homogeneous differential equation.

(2) If μ is such that $(m - 1)m = \mu^2$ for some integer m, then a_{m+1} has to be 0 as must $a_{m+3}, a_{m+5}, \cdots$.

(3) If the series is not finite (finite meaning all coefficients past a certain one are 0) it will not converge, because a_m will not approach 0 as m increases.

3. For $P_5(x)$, $\mu^2 = 5 \cdot 6 = 30$. The coefficients are

$$a_3 = \frac{2 - 30}{6}a_1 = -\frac{14}{3}a_1, \quad a_5 = \frac{12 - 30}{20}a_3 = \frac{-9}{10}a_3 = \frac{21}{5}a_1.$$

Using a common denominator of 15, we find

$$P_5(x) = \frac{63x^5 - 70x^3 + 15x}{15}a_1.$$

The normalization condition is $P_5(1) = 1$, which becomes $\frac{8}{15}a_1 = 1$, so $P_5(x) = (63x^5 - 70x^3 + 15x)/8$.

From Equation (9) with $n = 4$, we get $5P_5(x) + 4P_3(x) = 9xP_4(x)$. Solving this for $P_5(x)$ and substituting known expressions for P_3 and P_4 gives

$$P_5(x) = \frac{9}{40}(35x^5 - 30x^3 + 3x) - \frac{2}{5}(5x^3 - 3x)$$

and algebraic reduction leads to the same result as above.

5. The method of Section 0-2B works; however, it is much easier to take the equation in the form given in the Summary, with $\mu = 0$; $((1 - x^2y')' = 0$. Then $y' = \dfrac{c}{(1 - x^2)}$, and $y_2(x) = \dfrac{c}{2}\ln\left(\left|\dfrac{1 + x}{1 - x}\right|\right)$ is a solution of the differential equation that is independent of $y_1(x) = 1$. Note that $y_2(x)$ is unbounded at $x = \pm 1$.

7. First, differentiate Equation (9) $(n + 1)P'_{n+1} = (2n + 1)(P_n + xP'_n) - nP'_{n-1}$. Replace P'_{n-1} by using this rearranged version of Equation (8): $P'_{n-1} = P'_{n+1} - (2n + 1)P_n$, to get

$$(n + 1)P'_{n+1} = (2n + 1)(P_n + xP'_n) - n(P'_{n+1} - (2n + 1)P_n)$$

$$(2n + 1)P'_{n+1} = (2n + 1)(P_n + xP'_n) + n(2n + 1)P_n$$

$$P'_{n+1} = (n + 1)P_n + xP'_n$$

9. To do the differentiation, you need the general product rule:

$$(fg)^{(k)} = \sum_{j=0}^{k} \binom{k}{j} f^{(j)} g^{(k-j)}.$$

Now, use $k = n + 1$, $(x^2 - 1) = f(x)$ and $F'(x) = g(x)$, noting that $f'''(x) \equiv 0$.

$$\left((x^2 - 1)F'\right)^{(n+1)} = (x^2 - 1)F^{(n+2)} + (n + 1) \cdot 2xF^{(n+1)} + (n + 1)nF^{(n)}.$$

Next, use $x = f(x)$ and $F(x) = g(x)$ to differentiate the right-hand side

$$2n(xF)^{(n+1)} = 2n(xF^{(n+1)} + nF^{(n)}).$$

The differentiated equation is

$$(x^2 - 1)F^{(n+2)} + (n + 1) \cdot 2xF^{(n+1)} + (n + 1)nF^{(n)} = 2nxF^{(n+1)} + 2n^2 F^{(n)}$$

or

$$(x^2 - 1)F^{(n+2)} + 2xF^{(n+1)} - n(n - 1)F^{(n)} = 0.$$

This is Equation (5) with $y = F^{(n)}$, $\mu^2 = n(n - 1)$.

11. The given function is even, so $b_n = 0$ for odd values of n, and $b_n = (2n + 1) \int_0^1 xP_n(x)dx$ for even values of n. From Equation (14),

$$\int_0^1 xP_n(x)dx = \frac{(1 - x^2)}{(n + 2)(n - 1)} \left(P_n(x) - xP_n'(x)\right)\Big|_0^1 = \frac{-1}{(n + 2)(n - 1)} P_n(0).$$

From Equation (15) we have, for $n = 2m$

$$b_{2m} = \frac{(4m + 1)}{(2m + 2)(2m - 1)}(-1)^{m+1}\frac{1 \cdot 3 \cdots (2m - 1)}{2 \cdot 4 \cdots (2m)}.$$

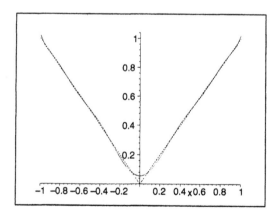

The figure shows the sum of terms up to $P_{10}(x)$.

Chapter 5

5.10 Some Applications of Legendre Polynomials

1. See Answers. Note that $u(1, \phi) - \frac{1}{2}$ is an odd function.

3. The "steady-state solution" is just $v(\phi) = T$ (constant). If we set $u(\phi, t) = T + w(\phi, t)$, then w satisfies

$$\frac{1}{\sin(\phi)} \frac{\partial}{\partial \phi} \left(\sin(\phi) \frac{\partial w}{\partial \phi} \right) = R^2 \left(\frac{1}{k} \frac{\partial w}{\partial t} + \gamma^2 w \right).$$

$$w(\phi, 0) = -T, \quad 0 < \phi < \pi.$$

Now proceed with separation of variables: $w(\phi, t) = \Phi(\phi) T(t)$

$$\frac{(\sin(\phi)\Phi')'}{\sin(\phi)\Phi} = R^2 \left(\frac{T'}{kT} + \gamma^2 \right) = -\mu^2.$$

Thus Φ satisfies the eigenvalue problem of Equation (4) – with boundedness conditions at $\phi = 0, \pi$, of course. We know that the solution is $\mu_n = n(n+1)$, $\Phi_n(\phi) = P_n(\cos(\phi))$, $n = 0, 1, 2, \cdots$. The problem for $T_n(t)$ is

$$T_n' + k \left(\gamma^2 + \frac{\mu_n^2}{R^2} \right) T_n = 0$$

with solution

$$T_n(t) = \exp \left(-k \left(\gamma^2 + \frac{\mu_n^2}{R} \right) t \right).$$

The form of the solution is

$$w(\phi, t) = \sum_{n=0}^{\infty} b_n P_n(\cos(\phi)) T_n(t)$$

and the b_n are determined by the initial condition $w(\phi, 0) = -T$, $0 < \phi < \pi$, or

$$\sum_{n=0}^{\infty} b_n P_n(\cos(\phi)) = -T, \quad 0 < \phi < \pi$$

so that

$$b_n = \frac{2n+1}{2} \int_0^\pi (-T) P_n(\cos(\phi)) \sin(\phi) d\phi.$$

As in part B, we may easily change variables to get

$$b_n = -T \frac{2n+1}{2} \int_{-1}^1 P_n(x) dx.$$

All of the coefficients are 0 except $b_0 = -T$. Thus the solution is

$$u(\phi, t) = T \left(1 - e^{-\gamma^2 kt} \right).$$

The usual procedures disguise the simplicity of the solution.

5. The solution follows part C, except that the condition $u(r, \frac{\pi}{2}) = 0$ forces odd indexes only:

$$\Phi_n(\phi) = P_{2n-1}(\cos(\phi)), \quad n = 1, 2, \cdots$$

$$\mu_n^2 = (2n - 1)2n$$

$$R_n(\rho) = \rho^{-\frac{1}{2}} J_{2n+\frac{1}{2}}(\lambda_{mn}\rho)$$

and λ_{mn} must be chosen to make $R_n(a) = 0$.

7. The nodal surfaces are places where the eigenfunction is 0. The factor of $\rho^{-\frac{1}{2}}$ plays no role here, so we need to have $J_{5/2}(\lambda\rho)P_2(\cos(\phi)) = 0$, where λ is the second positive solution of $J_{5/2}(\lambda) = 0$.

Now $J_{5/2}(\lambda) = \lambda^{-3}[(3 - \lambda^2)\sin(\lambda) - 3\lambda\cos(\lambda)]$. The solutions must be found numerically: $\lambda_1 = 5.76$, $\lambda_2 = 9.095$, $\lambda_3 = 12.32$, $\lambda_4 = 15.51$, etc. For $\lambda = \lambda_2$, $J_{5/2}(\lambda\rho) = 0$ at $\rho = \lambda_1/\lambda_2 = .633$ and of course at $\rho = 0$ and $\rho = 1$.

The function $P_2(x) = 0$ at $x = \pm 1/\sqrt{3}$, so $P_2(\cos(\phi)) = 0$ at $\phi = \cos^{-1}(\pm 1/\sqrt{3}) = .955$, $\pi - .955$. Thus, the nodal surface consists of the cones $\phi = .955$ and $\phi = \pi - .955$ and the spheres $\rho = .633$ and $\rho = 1$. See the Figures.

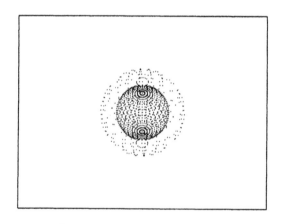

Exercise 7 Two spheres

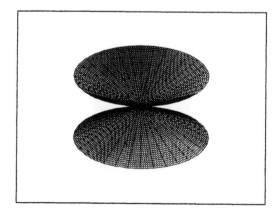

Exercise 7 Two cones

9. To determine the differential equation for u, use the chain rule:

$$\frac{\partial T}{\partial \rho} = \frac{\partial u}{\partial r}\frac{dr}{d\rho} = \frac{1}{R}\frac{\partial u}{\partial r}.$$

The original partial differential equation becomes

$$\frac{1}{(Rr)^2}\left[\frac{1}{R}\frac{\partial}{\partial r}\left((rR)^2\frac{1}{R}\frac{\partial u}{\partial r}\right) + \frac{1}{\sin(\phi)}\frac{\partial}{\partial \phi}\left(\sin(\phi)\frac{\partial u}{\partial \phi}\right)\right] = 0.$$

The Rr outside can be dropped; R inside cancels, leaving the equation cited. At the boundary $\phi = \pi/2$,

$$u\left(r, \frac{\pi}{2}\right) + T_W = T_W, \quad 0 < r < 1.$$

Of course, T_w drops out. At $\rho = R$ or $r = 1$,

$$\frac{k}{R}\frac{\partial u}{\partial r}(1, \phi) = h\left[T_G - (u(1, \phi) + T_W)\right]$$

or

$$\frac{k}{Rh}\frac{\partial u}{\partial r}(1, \phi) + u(1, \phi) = T_G - T_W, \quad 0 < \phi < \pi/2.$$

Chapter 5

Miscellaneous

1. Follow the development in the text, except that: $\mu_0^2 = 0$, $X_0(x) = 1$, and $\mu_m^2 = (m\pi/a)^2$, $X_m(x) = \cos(m\pi x/a)$. The form of the solution is

$$u(x, y, t) = \sum_{n=1}^{\infty} a_{0n} \sin\left(\frac{n\pi y}{b}\right) e^{-\nu_n^2 kt} + \sum_{m=1}^{\infty} \sum_{n=1}^{\infty} a_{mn} \cos\left(\frac{m\pi x}{a}\right) \sin\left(\frac{n\pi y}{b}\right) e^{-\lambda_{mn}^2 kt}.$$

The coefficients are found from the initial condition $u(x, y, 0) = Tx/a$.

$$a_{0n} = \frac{2}{ab} \int_0^a \int_0^b \frac{Tx}{a} \sin\left(\frac{n\pi y}{b}\right) dy dx$$

$$a_{mn} = \frac{4}{ab} \int_0^a \int_0^b \frac{Tx}{a} \cos\left(\frac{m\pi x}{a}\right) \sin\left(\frac{n\pi y}{b}\right) dy dx.$$

Since the region is a rectangle and the integrand is a product of a function of x by one of y, the iterated integral becomes a product, and

$$a_{0n} = T\frac{1 - \cos(n\pi)}{n\pi}$$

$$a_{mn} = -4T\frac{1 - \cos(m\pi)}{m^2\pi^2} \frac{1 - \cos(n\pi)}{n\pi}.$$

3. The solution follows the text down through Equation (16). The integration for coefficients is

$$a_{mn} = \frac{4}{ab} \int_0^b \int_0^a T\sin\left(\frac{m\pi x}{a}\right) \sin\left(\frac{n\pi y}{b}\right) dx dy = 4T\frac{1 - \cos(m\pi)}{m\pi} \cdot \frac{1 - \cos(n\pi)}{n\pi}.$$

The first three nonzero terms are $(m, n) = (1, 1)$, $(1,3)$, $(3,1)$, $(3,3)$

$$a_{11} = \frac{16T}{\pi^2}, \quad a_{13} = a_{31} = \frac{16T}{3\pi^2}, \quad a_{33} = \frac{16T}{a\pi^2}.$$

The exponent multipliers are, for $a = b$,

$$\lambda_{11}^2 k = \left(\frac{\pi}{a}\right)^2 2k; \quad \lambda_{13}^2 k = \lambda_{31}^2 k = \left(\frac{\pi}{a}\right)^2 \cdot 10k, \quad \lambda_{33}^2 k = \left(\frac{\pi}{a}\right)^2 18k.$$

See Answers.

5. Solve directly by integration (See Chapter 0)

$$u = -\frac{r^2}{2} + c_1 \ln(r) + c_2.$$

Apply boundedness ($c_1 = 0$) and boundary ($c_2 = a^2/2$) conditions:

$$u(r) = \frac{(a^2 - r^2)}{2}.$$

To solve by Bessel series, recall that $R = J_0(\lambda r)$ is the bounded solution of $(rR')' + \lambda^2 rR = 0$. Thus,

$$\frac{1}{r}(ru')' = \sum_{n=1}^{\infty} C_n \frac{1}{r}\frac{d}{dr}\left(r\frac{dJ_0(\lambda_n r)}{dr}\right)$$

$$= \sum_{n=1}^{\infty} C_n(-\lambda_n^2) J_0(\lambda_n r)$$

$$= \sum_{n=1}^{\infty} c_n J_0(\lambda_n r).$$

Match coefficients to determine

$$C_n = -\frac{c_n}{\lambda_n^2}.$$

From Section 5-6, Equation (17), we know that $c_n = \dfrac{1}{\alpha_n J_1(\alpha_n)}$ where α_n is the nth positive solution of $J_0(x) = 0$ and $\lambda_n = \alpha_n/a$.

7. This idea works for Exercise 1 because the initial condition is a function of x alone. The functions w and v in Exercise 6 satisfy the heat equation and these conditions

$$\frac{\partial w}{\partial x}(0,t) = 0, \quad \frac{\partial w}{\partial x}(a,t) = 0 \qquad\qquad v(0,t) = 0, \quad v(b,t) = 0$$

$$w(x,0) = \frac{Tx}{a}, \quad 0 < x < a \qquad\qquad v(y,0) = 1, \quad 0 < y < b$$

By methods of Chapter 2, we find

$$w(x,t) = T\left(\frac{1}{2} + \sum_{m=1}^{\infty} 2\frac{\cos(m\pi) - 1}{m^2\pi^2}\cos(\mu_m x)e^{-\mu_m^2 kt}\right)$$

with $\mu_m = m\pi/a$.

$$v(y,t) = \sum_{n=1}^{\infty} 2\frac{1 - \cos(n\pi)}{n\pi}\sin(\nu_n y)e^{-\nu_n^2 kt}$$

with $\nu_n = n\pi/b$. Then the solution of Exercise 1 is just $u(x,y,t) = w(x,t)v(y,t)$.

9. Separation of variables proceeds as in Section 5-6. Since $J_0(\lambda r)$ is indeed bounded as $r \to \infty$ (never stated, but suggested by Figure 7) the product solutions are

$$J_0(\lambda r)e^{-\lambda^2 kt}, \quad 0 \le \lambda.$$

11. Because $((1 - x^2)P_k')' = -k(k+1)P_k$, the differential equation becomes

$$\sum_{k=0}^{\infty} -k(k+1)B_k P_k(x) = -\sum_{k=0}^{\infty} b_k P_k(x)$$

and hence $B_k = b_k/(k(k+1))$. Note that $P_0(x)$ disappears on the left, so b_0 must be 0, or no solution exists. That is the content of the last statement in Exercise 10.

13. Follow the change of variable in Section 5-9 from Equation (4) to (5). See Answers.

15. The separation of variables follows Section 5.8A, except that the boundary condition in Equation (7) is replaced by $R(a) = 0$. The product solutions are

$$u_n(r, z) = J_0(\lambda_n r) \sinh(\lambda_n z)$$

with $\lambda_n = \alpha_n/a$ and α_n is the nth positive solution of $J_0(x) = 0$. (There is also a solution with $\cosh(\lambda_n z)$, but it is eliminated by the condition $u(r, 0) = 0$.) A convenient form for the solution is

$$u(r, z) = \sum_{n=1}^{\infty} a_n J_0(\lambda_n r) \frac{\sinh(\lambda_n z)}{\sinh(\lambda_n b)}$$

Then, the condition at $z = b$ becomes

$$\sum_{n=1}^{\infty} a_n J_0(\lambda_n r) = U_0, \quad 0 < r < a.$$

The coefficients were found in Section 5-6:

$$a_n = \frac{2U_0}{\alpha_n J_1(\alpha_n)}.$$

17. Separate variables: $u(r, z, t) = R(r)Z(z)T(t)$. The partial differential equation becomes

$$\frac{(rR')'}{rR} + \frac{Z''}{Z} = \frac{T''}{c^2 T}.$$

Eigenvalue problems are

$$(rR')' + \lambda^2 rR = 0,$$
$$R(0) \text{ bounded}, \quad R(a) = 0$$
$$Z'' + \mu^2 Z = 0,$$
$$Z(0) = 0, \quad Z(b) = 0.$$

Both are now familiar, so we find $R_n(r) = J_0(\lambda_n r)$, $\lambda_n = \alpha_n/a$; $Z_m(z) = \sin(\mu_m z)$, $\mu_m = m\pi/b$. The equation for T_{mn} is $T''_{mn} + c^2(\mu_m^2 + \lambda_n^2)^2 T_{mn} = 0$. Thus the frequencies are $c\sqrt{\mu_m^2 + \lambda_n^2}$ (in rad/sec – divide by 2π for Hz).

19. See Answers and the Figure.

21. See Answers and the Figure.

23. For the equilateral triangle with base $y = 0$, $0 < x < 1$, the lowest eigenvalue is $\lambda_1^2 = 16\pi^2/3$ (from Exercise 21). For one-sixth of a circular disk, the eigenvalue problem is

$$\frac{1}{r} \frac{\partial}{\partial r} \left(r \frac{\partial \phi}{\partial r} \right) + \frac{1}{r^2} \frac{\partial^2 \phi}{\partial \theta^2} = -\lambda^2 \phi, \quad 0 < r < 1, \quad 0 < \theta < \frac{\pi}{3}$$

with boundary conditions

$$\phi(1, \theta) = 0, \quad \phi(r, 0) = 0, \quad \phi\left(r, \frac{\pi}{3}\right) = 0.$$

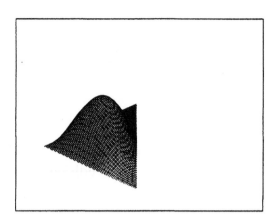

Exercise 19: First eigenfunction of isosceles right triangle

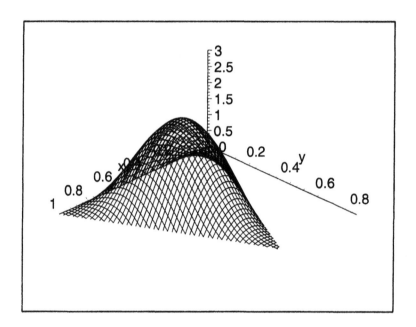

Exercise 21: First eigenfunction of equilateral triangle

Use the product form $\phi(r, \theta) = R(r)\Theta(\theta)$ to separate variables. The problem becomes

$$\frac{(rR')'}{rR} + \frac{1}{r^2}\frac{\Theta''}{\Theta} = -\lambda^2$$

and the separate eigenvalue problems are

$$\Theta'' + \mu^2\Theta = 0, \quad \Theta(0) = 0, \quad \Theta\left(\frac{\pi}{3}\right) = 0$$

with solution $\Theta_m = \sin(\mu_m\theta)$, $\mu_m = 3m$, $m = 1, 2, \cdots$.

$$(rR')' - \frac{\mu^2}{r}R + \lambda^2 rR = 0, \quad R(1) = 0$$

(see Section 5-5), with solution (bounded at $r = 0$)

$$R_{mn}(r) = J_{3m}(\lambda_{mn}r)$$

where $\lambda_{mn} = \alpha_{3m,n}$ is the nth positive solution of $J_{3m}(x) = 0$. Thus the least eigenvalue for this region is $\alpha_{31} = 6.38$, somewhat less than the least eigenvalue of an equilateral triangle, $\pi\sqrt{16/3} = 7.26$.

25. $\nabla^2 u = \dfrac{1}{r}\dfrac{\partial}{\partial r}\left(r\dfrac{\partial u}{\partial r}\right)$ for a function that depends on the polar coordinate r but not on θ or z. For the function given, compute

$$u_r = t^b\left(-\frac{2r}{4t}\right)e^{\frac{-r^2}{4t}} = -\frac{r}{2t}u$$

$$ru_r = -\frac{r^2}{2t}u$$

$$(ru_r)_r = -\frac{r}{t}u - \frac{r^2}{2t}\left(\frac{-r}{2t}u\right) = \left(-\frac{r}{t} + \frac{r^3}{4t^2}\right)u$$

$$\frac{1}{r}(ru_r)_r = \left(-\frac{1}{t} + \frac{r^2}{4t^2}\right)u$$

$$u_t = bt^{b-1}e^{-\frac{r^2}{4t}} + t^b\left(+\frac{r^2}{rt^2}\right)e^{-\frac{r^2}{4t}} = \left(\frac{b}{t} + \frac{r^2}{4t^2}\right)u.$$

By matching, you see that $b = -1$.

27. The function y must satisfy $(xy')' + k^2 xy = 0$, $y(0)$ bounded, $y(a) = h$, with $k^2 = \omega^2/gU$. The bounded solution of the differential equation is $y(x) = cJ_0(kx)$. The boundary condition requires $cJ_0(ka) = h$, which fails to have a solution if $J_0(ka) = 0$. In this case, resonance occurs, and the form suggested is not adequate.

29. Assume $u(r, t) = R(r)T(t)$. Then

$$\left(r^{n-1}R'\right)' + \lambda^2 r^{n-1}R = 0, \quad T'' + c^2\lambda^2 T = 0.$$

The equation for R can be solved by comparing to Equation (1) of Section 5-8, giving $\alpha = p = (2-n)/2$, $\gamma = 1$. The bounded solution is $R(r) = r^\alpha J_\alpha(\lambda r)$.

31. In the change of variables, use the chain rule to find derivatives

$$\frac{\partial C}{\partial r} = \frac{\partial u}{\partial \rho}\frac{\partial \rho}{\partial r} = \frac{\partial u}{\partial \rho}\cdot\frac{1}{R}, \quad \frac{\partial C}{\partial z} = \frac{\partial u}{\partial \zeta}\frac{\partial \zeta}{\partial z} = \frac{\partial u}{\partial \zeta}\cdot\frac{1}{L},$$

etc. The equation becomes

$$\frac{D_r}{\rho R}\frac{\partial}{\partial \rho}\left(\rho R\frac{\partial u}{\partial \rho}\frac{1}{R}\right)\cdot\frac{1}{R} + \frac{C_z}{L^2}\frac{\partial^2 u}{\partial \zeta^2} + \frac{U}{L}\frac{\partial u}{\partial \zeta} = 0.$$

Multiply through by L^2/D_z. Then

$$b = \frac{D_r L^2}{D_z R^2}, \quad p = \frac{UL}{D_z}.$$

33. The trapezoidal rule approximation for an integral over the interval 0 to 1, with 10 subintervals ($\Delta x = .1$) is

$$\int_0^1 g(x)dx \simeq \frac{1}{10}\left[\frac{1}{2}g(0) + g(.1) + \cdots + g(.9) + \frac{1}{2}g(1)\right].$$

In the table the information for computing a_0 and a_1 is set up. In columns 3-6, the w stands for weight: 1/2 at $\rho = 0$ and $\rho = 1$, 1 between. The last row is the sum of the numbers above it, multiplied by 1/10.

Using the information in the table, the required approximations are $a_0 = 2 \cdot 6.386 = 12.77$, $a_1 = -0.397/0.0801 = -4.91$.

35. Separation of variables leads to

$$\frac{(r^2 R')'}{r^2 R} = \frac{T'}{T} = -\lambda^2$$

The boundary condition becomes $R(1)T'(t) = -DR'(1)T(t)$. Since $T'/T = -\lambda^2$, this becomes

$$DR'(1) = \lambda^2 R(1)$$

The differential equation $(r^2 R')' + \lambda^2 r^2 R = 0$ has the bounded solution $R(r) = \sin(\lambda r)/r$. The boundary condition is

$$D(\lambda \cos(\lambda) - \sin(\lambda)) = \lambda^2 \sin(\lambda)$$

or

$$\tan(\lambda) = D\lambda/(\lambda^2 + D).$$

The first positive solution is $\phi(D = 0)$, $3.173(D = 1)$, $4.132(D = 10)$. Also $\lambda = 0$ corresponds to $R(r) = 1$.

Chapter 6

6.1 Definition and Elementary Properties

1.a. $\sinh(at) = \dfrac{1}{2}\left(e^{at} - e^{-at}\right)$

$$\mathcal{L}\left(\sinh(at)\right) = \frac{1}{2}\left(\frac{1}{s-a} - \frac{1}{s+a}\right) = \frac{a}{s^2 - a^2}$$

b. $\cos(\omega t) = \dfrac{1}{2}\left(e^{i\omega t} + e^{-i\omega t}\right)$

$$\mathcal{L}\left(\cos(\omega t)\right) = \frac{1}{2}\left(\frac{1}{s-i\omega} + \frac{1}{s+i\omega}\right) = \frac{s}{s^2 + \omega^2}$$

c. $\cos^2(\omega t) = \dfrac{1}{4}\left(e^{2i\omega t} + 2 + e^{-2i\omega t}\right)$

$$\mathcal{L}\left(\cos^2(\omega t)\right) = \frac{1}{4}\left(\frac{1}{s - 2i\omega} + \frac{2}{s} + \frac{1}{s + 2i\omega}\right) = \frac{1}{4}\left(\frac{2s}{s^2 + 4\omega^2} + \frac{2}{s}\right) = \frac{s^2 + 2\omega^2}{s(s^2 + 4\omega^2)}$$

d. $\sin(\omega t - \phi) = \dfrac{1}{2i}\left(e^{i(\omega t - \phi)} - e^{-i(\omega t - \phi)}\right)$

$$\mathcal{L}\left(\sin(\omega t - \phi)\right) = \frac{1}{2i}\left(\frac{e^{-i\phi}}{s - i\omega} - \frac{e^{i\phi}}{s + i\omega}\right) = \frac{1}{2i}\frac{s\left(e^{-i\phi} - e^{i\phi}\right) + i\omega\left(e^{-i\phi} - e^{i\phi}\right)}{s^2 + \omega^2}$$

$$= \frac{1}{s^2 + \omega^2}\left[s\frac{e^{-i\phi} - e^{i\phi}}{2i} + \omega\frac{e^{-i\phi} + e^{i\phi}}{2}\right] = \frac{-s\sin(\phi) + \omega\cos(\phi)}{s^2 + w^2}$$

e. $e^{2(t+1)} = e^2 \cdot e^{2t}; \quad \mathcal{L}\left(e^{2(t+1)}\right) = \dfrac{e^2}{s - 2}$

f. $\sin^2(\omega t) = 1 - \cos^2(\omega t)$ (see part c.)

3.a. $\mathcal{L}(f(t)) = \displaystyle\int_a^\infty e^{-st}\,dt = \left.\frac{e^{-st}}{-s}\right|_a^\infty = \frac{e^{-as}}{s}$

b. $\mathcal{L}(f(t)) = \displaystyle\int_a^b e^{-st}\,dt = \left.\frac{e^{-st}}{-s}\right|_a^b = \frac{e^{-as} - e^{-bs}}{s}$

c. $\mathcal{L}(f(t)) = \displaystyle\int_0^a e^{-st} \cdot t\,dt + \int_a^\infty e^{-st} \cdot a\,dt$

$$= \left.\frac{-st - 1}{s^2}e^{-st}\right|_0^a + \left.\frac{e^{-st}}{-s}\right|_a^\infty$$

$$= \frac{-sa - 1}{s^2}e^{-sa} + \frac{1}{s^2} + \frac{e^{-sa}}{s}$$

$$= \frac{1 - e^{-sa}}{s^2}$$

5.a $\dfrac{1}{s^2 + 2s} = \dfrac{1}{s^2 + 2s + 1 - 1} = \dfrac{1}{(s+1)^2 - 1} = F(s + 1)$

$$F(s) = \frac{1}{s^2 - 1}; \quad \mathcal{L}^{-1}\left(\frac{1}{s^2 - 1}\right) = \sinh(t)$$

$$\mathcal{L}^1\left(\frac{1}{s^2 + 2s}\right) = e^{-t}\sinh(t) = \frac{1}{2}\left(1 - e^{-2t}\right)$$

b. $\dfrac{s + 1}{s^2 + 2s + 2} = \dfrac{s + 1}{s^2 + 2s + 1 + 1} = \dfrac{s + 1}{(s + 1)^2 + 1} = F(s + 1)$

$$F(s) = \frac{s}{s^2 + 1}; \quad \mathcal{L}^{-1}\left(\frac{s}{s^2 + 1}\right) = \cos(t)$$

$$\mathcal{L}^{-1}\left(\frac{s + 1}{s^2 + 2s + 2}\right) = e^{-t}\cos(t)$$

c. Denominator $s^2 + 2as + b^2 = s^2 + 2as + a^2 + b^2 - a^2 = (s + a)^2 + (b^2 - a^2)$. Let $b^2 - a^2 = \omega^2$.

$$\frac{1}{s^2 + 2as + b^2} = \frac{1}{(s + a)^2 + \omega^2} = F(s + a)$$

$$F(s) = \frac{1}{s^2 + \omega^2}; \quad \mathcal{L}^{-1}\left(\frac{1}{s^2 + \omega^2}\right) = \frac{\sin(\omega t)}{\omega}$$

$$\mathcal{L}^{-1}\left(\frac{1}{s^2 + 2as + b^2}\right) = e^{-at}\frac{\sin(\omega t)}{\omega}$$

7.a. Use partial fractions: $\dfrac{1}{(s - a)(s - b)} = \dfrac{1}{a - b}\left(\dfrac{1}{s - a} - \dfrac{1}{s - b}\right)$

$$\mathcal{L}^{-1}\left(\frac{1}{(s - a)(s - b)}\right) = \frac{1}{a - b}\left(e^{at} - e^{bt}\right).$$

b. The squared denominator suggests that the function is a derivative. Check:

$$\frac{d}{ds}\left(\frac{1}{s^2 - a^2}\right) = \frac{-2s}{(s^2 - a^2)^2}. \quad \text{Thus} \quad \frac{s}{(s^2 - a^2)^2} = -\frac{1}{2}\frac{d}{ds}\left(\frac{1}{s^2 - a^2}\right).$$

Knowing $\dfrac{a}{s^2 - a^2} = \mathcal{L}(\sinh(at))$ and using Equation (6), $\dfrac{s}{(s^2 - a^2)^2} = \dfrac{1}{2}\mathcal{L}\left(\dfrac{t}{a}\sinh(at)\right).$

c. Similar to part b. $\dfrac{d}{ds}\left(\dfrac{1}{s^2 + \omega^2}\right) = \dfrac{-2s}{(s^2 + \omega^2)^2},$ so $\dfrac{s^2}{(s^2 + \omega^2)^2} = -\dfrac{s}{2}\dfrac{d}{ds}\left(\dfrac{1}{s^2 + \omega^2}\right)$

$$-\frac{d}{ds}\left(\frac{1}{s^2 + \omega^2}\right) = \mathcal{L}\left(t\frac{\sin(\omega t)}{\omega}\right), \quad \text{so} \quad -\frac{s}{2}\frac{d}{ds}\left(\frac{1}{s^2 + \omega^2}\right) = \mathcal{L}\left(\frac{d}{dt}\left(\frac{t\sin(\omega t)}{2\omega}\right)\right)$$

$$= \frac{1}{2\omega}\left(\omega t \cos(\omega t) + \sin(\omega t)\right). \text{ Also see Table 2.}$$

d. Use shifting. $\dfrac{1}{(s - a)^3} = F(s - a); \quad f(s) = \dfrac{1}{s^3}, \quad \mathcal{L}^{-1}\left(\dfrac{1}{s^3}\right) = \dfrac{t^2}{2}$ (from the text). Therefore $\mathcal{L}^{-1}\left(\dfrac{1}{(s - a)^3}\right) = \dfrac{e^{at}t^2}{2}.$

e. See Exercise 3b: use $a = 0$, $b = 1$

$$\mathcal{L}^{-1}\left(\frac{1 - e^{-s}}{s}\right) = \begin{cases} 1, & 0 < t < 1 \\ \\ 0, & 1 < t \end{cases}$$

9. Do these exercises starting from **c.**.

c. This is the same as 7c:

$$F(s) = \frac{s^2}{(s^2 + \omega^2)^2}, \, f(t) = \frac{1}{2\omega}(\omega t \cos(\omega t) + \sin(\omega t))$$

d. The function is $sF(s)$. Since $f(0) = 0$,

$$\mathcal{L}^{-1}(sF(s)) = f'(t) = \frac{1}{2}(\cos(\omega t) - \omega t \sin(\omega t))$$

b. The function is $\frac{1}{s}F(s)$. By the integral property, Eq. (5),

$$\mathcal{L}^{-1}(\frac{1}{s}F(s)) = \int_0^t f(t')dt' = \frac{1}{2\omega}\int_0^t [\omega t' \cos(\omega t') + \sin(\omega t')]dt'$$

$$= \frac{1}{2\omega}t \sin(\omega t)$$

a. Use Eq. (5) again:

$$\frac{1}{(s^2 + \omega^2)^2} = \int_0^t \frac{1}{2\omega}t' \sin(\omega t')dt' = \frac{1}{2\omega^2}[\sin(\omega t) - \omega t \cos(\omega t)]$$

Chapter 6

6.2 Partial Fractions and Convolutions

In these exercises, we use $\hat{u} = \mathcal{L}(u)$ etc. In 1 and 3, we give the transformed differential equation, $\hat{u}(s)$, then the inversion of $\hat{u}(s)$.

1.a. $s\hat{u} - 1 - 2\hat{u} = 0;$ $\quad \hat{u} = \dfrac{1}{s-2};$ $\quad u(t) = e^{2t}$

b. $s\hat{u} - 1 + 2\hat{u} = 0;$ $\quad \hat{u} = \dfrac{1}{s+2};$ $\quad u(t) = e^{-2t}$

c. $s^2\hat{u} - s + 4(s\hat{u} - 1) + 3\hat{u} = 0;$ $\quad \hat{u} = \dfrac{4+s}{s^2 + 4s + 3} = \dfrac{A}{s+1} + \dfrac{B}{s+3};$

$A = \dfrac{4-1}{2(-1)+4} = \dfrac{3}{2},$ $\quad B = \dfrac{4-3}{2(-3)+4} = -\dfrac{1}{2}$ (Heaviside formula);

$u(t) = \dfrac{3}{2}e^{-t} - \dfrac{1}{2}e^{-3t}.$

d. $s^2\hat{u} - 1 + 9\hat{u} = 0;$ $\quad \hat{u} = \dfrac{1}{s^2 + 9};$ $\quad u(t) = \dfrac{1}{3}\sin(3t)$ (lookup).

3.a. $s\hat{u} + a\hat{u} = \dfrac{1}{s};$ $\quad \hat{u} = \dfrac{1}{s(s+a)} = \dfrac{1}{a}\left(\dfrac{1}{s} - \dfrac{1}{s+a}\right);$ $\quad u(t) = \dfrac{1}{a}\left(1 - e^{-at}\right).$

b. $s^2\hat{u} + \hat{u} = \dfrac{1}{s^2};$ $\quad \hat{u} = \dfrac{1}{s^2(s^2+1)} = \dfrac{1}{s^2} - \dfrac{1}{s^2+1}$ (by partial fractions). $u(t) = t - \sin(t).$

c. $s^2\hat{u} + 4\hat{u} = \dfrac{1}{s^2+1};$ $\quad \hat{u} = \dfrac{1}{(s^2+1)(s^2+4)} = \dfrac{1}{3}\left(\dfrac{1}{s^2+1} - \dfrac{1}{s^2+4}\right)$ (by partial fractions).

$u(t) = \dfrac{1}{3}\left(\sin(t) - \dfrac{1}{2}\sin(2t)\right).$

d. $s^2\hat{u} + 4\hat{u} = \dfrac{1}{(s^2+4)};$ $\quad \hat{u} = \dfrac{1}{(s^2+4)^2}.$ From the text Section 6.1,

$\mathcal{L}(t\sin(\omega t)) = 2\omega\dfrac{s}{(s^2+\omega^2)^2}.$ Now use the transform of the integral:

$$\mathcal{L}\left(\int_0^t \tau\sin(\omega\tau)d\tau\right) = \dfrac{1}{s}\mathcal{L}\left(t\sin(\omega t)\right) = \dfrac{2\omega}{(s^2+\omega^2)^2}.$$

Finally,

$$u(t) = \dfrac{1}{2\omega}\int_0^t \tau\sin(\omega\tau)d\tau = \dfrac{1}{2\omega^3}\left(\sin(\omega t) - \omega t\cos(\omega t)\right)$$

Now use $\omega = 2.$

e. $s^2\hat{u} + 2s\hat{u} = \dfrac{1}{s} - \dfrac{1}{s+1};$ $\quad \hat{u} = \dfrac{1}{s^2(s+1)(s+2)}.$ By partial fractions, you can find

$$\hat{u} = \dfrac{1}{2s^2} - \dfrac{3}{4s} + \dfrac{1}{s+1} - \dfrac{1}{4(s+2)} \quad \text{and} \quad u(t) = \dfrac{1}{2}t - \dfrac{3}{4} + e^{-t} - \dfrac{1}{4}e^{-2t}.$$

Alternatively, resolve

$$\frac{1}{s(s+1)(s+2)} = \frac{1}{2s} - \frac{1}{s+1} + \frac{1}{2(s+2)},$$

and use integration to find

$$u(t) = \int_0^t \left[\frac{1}{2} - e^{-\tau} + \frac{1}{2}e^{-2\tau} \right] d\tau = \frac{1}{2}t - \left(1 - e^{-t}\right) + \frac{1}{4}\left(1 - e^{-2t}\right)$$

$$= \frac{1}{2}t - \frac{3}{4} + e^{-t} - \frac{1}{4}e^{-2t}$$

f. $s^2\hat{u} - \hat{u} = \frac{1}{s}$; $\hat{u} = \frac{1}{s(s^2 - 1)} = -\frac{1}{s} + \frac{1}{2(s+1)} + \frac{1}{2(s-1)}$ by Heaviside's formula,

$$u(t) = -1 + \frac{1}{2}e^t + \frac{1}{2}e^{-t} = \cosh(t) - 1.$$

5.a. $\dfrac{1}{s^2 - 4} = \dfrac{A}{s-2} + \dfrac{B}{s+2}$; $A = \dfrac{1}{4}$, $B = -\dfrac{1}{4}$.

$$\mathcal{L}^{-1}\left(\frac{1}{s^2 - 4}\right) = \frac{1}{4}e^{2t} - \frac{1}{4}e^{-2t} = \frac{1}{2}\sinh(2t).$$

b. $\dfrac{1}{s^2 + 4} = \dfrac{A}{s-2i} + \dfrac{B}{s+2i}$; $A = \dfrac{1}{4i}$, $B = -\dfrac{1}{4i}$.

$$\mathcal{L}^{-1}\left(\frac{1}{s^2 + 4}\right) = \frac{1}{4i}\left(e^{2it} - e^{-2it}\right) = \frac{1}{2}\sin(2t).$$

c. $\dfrac{s+3}{s(s^2 + 2)} = \dfrac{A}{s} + \dfrac{B}{s - i\sqrt{2}} + \dfrac{C}{s + i\sqrt{2}}.$

$$A = \frac{3}{2}, \quad B = -\frac{(3 + i\sqrt{2})}{4}, \quad C = -\frac{(3 - i\sqrt{2})}{4}.$$

$$\mathcal{L}^{-1}\left(\frac{s+3}{s(s^2 + 2)}\right) = \frac{3}{2} - \frac{3}{4}\left(e^{i\sqrt{2}t} + e^{-i\sqrt{2}t}\right) + \frac{-i\sqrt{2}}{4}\left(e^{i\sqrt{2}t} - e^{-i\sqrt{2}t}\right)$$

Alternate:

$$\frac{s+3}{s(s^2 + 2)} = \frac{A}{s} + \frac{Bs + C}{s^2 + 2}, \qquad A = \frac{3}{2}, \quad B = -\frac{3}{2}, \quad C = 1$$

$$\mathcal{L}^{-1}\left(\frac{s+3}{s(s^2 + 2)}\right) = \frac{3}{2} - \frac{3}{2}\cos(\sqrt{2}t) + \frac{1}{\sqrt{2}}\sin(\sqrt{2}t).$$

d. $\dfrac{4}{s(s+1)} = \dfrac{A}{s} + \dfrac{B}{s+1}$; $A = 4$, $B = -4$, $\mathcal{L}^{-1}\left(\dfrac{4}{s(s+1)}\right) = 4 - 4e^{-t}.$

7.a. $f * g = \displaystyle\int_0^t 1 \cdot \sin(t')dt' = 1 - \cos(t).$ Check:

$$\frac{1}{s} \cdot \frac{1}{s^2 + 1} = \frac{1}{s} - \frac{s}{s^2 + 1}.$$

b. $f * g = \int_0^t e^{(t-t')} \cos(\omega t') dt' = e^t \int_0^t e^{-t'} \cos(\omega t') dt' = e^t \cdot e^{-t'} \left. \dfrac{-\cos(\omega t') + \omega \sin(\omega t')}{\omega^2 + 1} \right|_0^t$

$$= e^t \left[e^{-t} \dfrac{-\cos(\omega t) + \omega \sin(\omega t)}{\omega^2 + 1} + \dfrac{1}{\omega^2 + 1} \right] = \dfrac{-\cos(\omega t) + \omega \sin(\omega t) + e^t}{\omega^2 + 1}.$$

Check:

$$\dfrac{1}{s-1} \cdot \dfrac{s}{s^2 + \omega^2} = \dfrac{A}{s-1} + \dfrac{Bs + C}{s^2 + \omega^2}, \quad A = -B = \dfrac{1}{1 + \omega^2}, \quad C = \dfrac{\omega^2}{1 + \omega^2}.$$

c. $f * g = \int_0^t (t - t') \sin(t') dt' = t \int_0^t \sin(t') dt' - \int_0^t t' \sin(t') dt'$

$$= t \left(-\cos(t') \Big|_0^t \right) - \left(\sin(t') - t' \cos(t') \Big|_0^t \right) = t(1 - \cos(t)) - (\sin(t) - t \cos(t)) = t - \sin(t)$$

Check:

$$\dfrac{1}{s^2} \cdot \dfrac{1}{s^2 + 1} = \dfrac{1}{s^2} - \dfrac{1}{s^2 + 1}$$

Chapter 6

6.3 Partial Differential Equations

It is useful to have these identities:

$\cosh(iy) = \cos(y)$, $\sinh(iy) = i\sin(y)$,
$\cosh(x + iy) = \cosh(x)\cos(y) + i\sinh(x)\sin(y)$,
$\sinh(x + iy) = \sinh(x)\cos(y) + i\cosh(x)\sin(y)$.

1.a. Let $\sqrt{s} = x + iy$, so we wish to have $\cosh(x + iy) = 0$. Then (see above) we must have $\cosh(x)\cos(y) = 0$ and $\sinh(x)\sin(y) = 0$. The cosh is never 0; thus $y = \pm(n - \frac{1}{2})\pi$ to make $\cos(y) = 0$, and then $x = 0$ to make $\sinh(x) = 0$. $\sqrt{s} = \pm i(n - \frac{1}{2})\pi$ and $s = -(n - \frac{1}{2})^2\pi^2$, $n = 1, 2, \cdots$.

b. From part **a.**, $s = \pm i(n - \frac{1}{2})\pi$.

c. Let $s = x + iy$. From above we must have $\sinh(x)\cos(y) = 0$, $\cosh(x)\sin(y) = 0$. Since $\cosh(x)$ is never 0, $y = \pm n\pi$, $n = 0, 1, 2, \cdots$; and then $\cos(y) = \pm 1$ for those values, so $x = 0$. Finally $s = \pm in\pi$, $n = 0, 1, 2, \cdots$.

d. $s = x + iy$. Then $\cosh(s) = s\sinh(s)$ becomes $\cosh(x)\cos(y) + i\sinh(x)\sin(y) = (x + iy)(\sinh(x)\cos(y) + i\cosh(x)\sin(y))$. Find the real and imaginary parts of the right-hand side and equate to the corresponding part of the left

$$\cosh(x)\cos(y) = x\sinh(x)\cos(y) - y\cosh(x)\sin(y) \tag{1}$$

$$\sinh(x)\sin(y) = x\cosh(x)\sin(y) + y\sinh(x)\cos(y). \tag{2}$$

Note that $\cos(y) = 0$ cannot satisfy (1), so divide by $\cosh(x)\cos(y)$ to find

$$1 + y\tan(y) = x\tanh(x). \tag{1'}$$

If $x = 0$, (2) is satisfied and $\tan(y) = -\frac{1}{y}$ has infinitely many solutions giving imaginary values of s. There are also two real solutions: $s = \pm x$, where $\tanh(x) = \frac{1}{x}$ or $x \cong 1.19968$.

e. Same procedure as above. Imaginary roots are $s = iy$ where $\tan(y) = \frac{1}{y}$. There is no real root.

3.a. Transform the partial differential equation (incorporating the initial condition) and boundary conditions:

$$U'' = sU, \quad 0 < x < 1$$

$$U(0) = 0, \quad U(1) = \frac{1}{s^2}.$$

Solution

$$U(x, s) = \frac{\sinh(\sqrt{s}x)}{s^2 \sinh(\sqrt{s})}$$

b.

$$U'' = sU - 1, \quad 0 < x < 1$$

$$U(0) = 0, \quad U(1) = \frac{1}{(s+1)}.$$

Solution:

$$U(x; s) = \frac{1}{s} + A\cosh(\sqrt{s}x) + B\sinh(\sqrt{s}x)$$

$$A = -\frac{1}{s}, \quad B = \frac{\frac{1}{s+1} - \frac{1}{s}(1 - \cosh\sqrt{s})}{\sinh(\sqrt{s})}.$$

The solution is much neater if we take

$$U(x; s) = \frac{1}{s} + C\frac{\sinh(\sqrt{s}x)}{\sinh(\sqrt{s})} + D\frac{\sinh(\sqrt{s}(1-x))}{\sinh(\sqrt{s})}.$$

Then $D = \dfrac{-1}{s}, \quad C = \dfrac{1}{s+1} - \dfrac{1}{s}.$

5.a.

$$U'' = sU, \quad U(0) = 0, \quad U(1) = \frac{1}{s}$$

$$U(x, s) = \frac{\sinh(\sqrt{s}x)}{s\sinh(\sqrt{s})}.$$

Follow Example 3. The denominator is zero at $s = 0$ and $s = -n^2\pi^2$, $n = 1, 2, \cdots$.

$$A_0 = x; \quad A_n = \frac{2\sin(n\pi x)}{n\pi \cos(n\pi)}; \quad \text{and} \quad u(x, t) = x + \sum_{n=1}^{\infty} \frac{2\sin(n\pi x)}{n\pi \cos(n\pi)}e^{-(n\pi)^2 t}.$$

b.

$$U'' = sU - 1, \quad U(0) = 0, \quad U(1) = 0$$

$$U(x, s) = \frac{1}{s} - \frac{\cosh(\sqrt{s}y)}{s\cosh(\sqrt{s}/2)},$$

where we have set $y = x - \frac{1}{2}$.

Now, $\mathcal{L}^{-1}(1/s) = 1$. Roots of the denominator of the second term of U occur at $s = 0$ and at $\sqrt{s} = i(2n-1)\pi$ or $s = -(2n-1)^2\pi^2$, $n = 1, 2, \cdots$.

At $s = 0$, $A_0 = 1$; at $\sqrt{s_n} = i(2n-1)\pi$, let

$q = \dfrac{\cosh(\sqrt{s}y)}{s}, \quad p = \cosh\left(\sqrt{s}/2\right), \quad p' = \dfrac{\sinh\left(\sqrt{s}/2\right)}{(4\sqrt{s})}.$ Then

$$\frac{q(s)}{p'(s)} = \frac{4\cosh(\sqrt{s}y)}{\sqrt{s}\sinh\left(\sqrt{s}/2\right)}$$

$$A_n(y) = \frac{4\cos((2n-1)\pi y)}{i(2n-1)\pi i\sin\left(\frac{(2n-1)\pi}{2}\right)}.$$

Now change back to x :

$$A_n(x) = -\frac{4\cos\left((2n-1)\pi(x-\frac{1}{2})\right)}{(2n-1)\pi\sin\left((n-\frac{1}{2})\pi\right)}$$

$$u(x,y) = 1 - 1 + \sum_{n=1}^{\infty}\frac{4\cos\left((2n-1)\pi(x-\frac{1}{2})\right)}{(2n-1)\pi\sin\left((n-\frac{1}{2})\pi\right)}e^{-(2n-1)^2\pi^2 t}.$$

Chapter 6

6.4 More Difficult Examples

1. Transformed problem:

$$U'' = sU, \quad U'(0) = 0, \quad U'(1) = \frac{1}{s}, \quad U(x, s) = \frac{\cosh(\sqrt{s}x)}{s\sqrt{s}\sinh(\sqrt{s})}.$$

Since we are seeking the persistent part, we are interested only in values of s *with non-negative real part* that make the denominator zero. The usual values of s that make $\sinh(\sqrt{s}) = 0$ have negative real parts. Therefore only $s = 0$ is of interest. At that point

$$s\sqrt{s}\sinh(\sqrt{s}) = s^{\frac{3}{2}}\left(\sqrt{s} + \frac{(\sqrt{s})^3}{6} + \cdots\right) = s^2 + \cdots$$

and

$$\cosh(\sqrt{s}x) = 1 + \frac{sx^2}{2} + \frac{s^2x^4}{24} + \cdots.$$

Thus, near $s = 0$, $U(x, s)$ behaves like

$$\frac{1 + \frac{sx^2}{2} + \frac{s^2x^4}{24} + \cdots}{s^2} = \frac{1}{s^2} + \frac{x^2}{2s}.$$

(The remaining terms do not go to $\pm\infty$ as $s \to 0$.) Thus, the persistent part of the solution is $u_\infty(x, t) = t + x^2/2$. It is easy to check that this function satisfies the heat equation and the boundary conditions.

3. Let $V(x, s)$ be the given function. By using Taylor series or L'Hopital's rule, you can see that V remains finite at $s = 0$. Therefore we want points where $\cosh(s/2) = 0$, which are $s_n = (2n-1)\pi i$, $s'_n = -(2n-1)\pi i$, $n = 1, 2, \cdots$. For determining $A_n(x)$, designate

$$q(s) = \frac{\cosh(s/2) - \cosh(sy)}{s^2},$$

with $y = x - 1/2$ for brevity.

$$p(s) = \cosh\left(\frac{s}{2}\right), \quad p'(s) = \frac{1}{2}\sinh\left(\frac{s}{2}\right).$$

Then $p'\left(\pm(2n-1)\pi i\right) = \pm\frac{i}{2}\sin\left((n - \frac{1}{2})\pi\right)$,

$$A_{\pm n} = \frac{2\cos((2n-1)\pi y)}{\pm i(2n-1)^2\pi^2\sin((n - \frac{1}{2})\pi)}$$

$$A_n e^{(2n-1)\pi it} + A_{-n}e^{-(2n-1)\pi it} = \frac{\cos((2n-1)\pi y)}{(2n-1)^2\pi^2\sin((n - \frac{1}{2})\pi)} \cdot 2\sin((2n-1)\pi t).$$

Finally

$$v(x, t) = \sum_{n=1}^{\infty} \frac{4}{(2n-1)^2\pi^2\sin\left((n - \frac{1}{2})\pi\right)}\cos\left((2n-1)\pi(x - \frac{1}{2})\right)\sin\left((2n-1)\pi t\right).$$

It is easy to see that $\dfrac{\partial^2 v}{\partial x^2} = \dfrac{\partial^2 v}{\partial t^2}$; $v(0, t) = v(1, t) = 0$; $v(x, 0) = 0$. The second initial condition is not obvious. Adding up some terms of the series shows that $v_t(x, 0) = 1$, $0 < x < 1$.

5.a. Transform the problem:

$$U'' = s^2 U - \sin(\pi x)\frac{\omega}{s^2 + \omega^2}$$

$$U(0) = 0, \quad U(1) = 0.$$

For this problem, the particular solution obtained by undetermined coefficients also satisfies the boundary conditions. The solution is thus

$$U(x, s) = \frac{\omega \sin(\pi x)}{(s^2 + \omega^2)(s^2 + \pi^2)}.$$

By partial fractions,

$$\frac{1}{(s^2 + \omega^2)(s^2 + \pi^2)} = \frac{1}{\pi^2 - \omega^2}\left(\frac{1}{s^2 + \omega^2} - \frac{1}{s^2 + \pi^2}\right),$$

so that

$$u(x, t) = \frac{\omega}{\pi^2 - \omega^2}\left(\frac{\sin(\omega t)}{\omega} - \frac{\sin(\pi t)}{\pi}\right)\sin(\pi x).$$

b. If we simply let $\omega = \pi$, we find

$$U(x, s) = \frac{\pi \sin(\pi x)}{(s^2 + \pi^2)^2},$$

and we can determine that (using Equation (5), Section 6.1)

$$\frac{2\pi}{(s^2 + \pi^2)^2} = \int_0^t t' \sin(\pi t')dt' = \left.\frac{\sin(\pi t')}{\pi^2} - \frac{t' \cos(\pi t')}{\pi}\right|_0^t = \frac{\sin(\pi t) - \pi t \cos(\pi t)}{\pi^2}.$$

Therefore

$$\mathcal{L}^{-1}(U(x, s)) = u(x, t) = \frac{1}{2\pi^2}\left(\sin(\pi t) - \pi t \cos(\pi t)\right)\sin(\pi x).$$

7. Transform the problem:

$$U'' = sU, \quad U(0) = 0, \quad U(1) = \frac{1}{s} - \frac{1}{s + a}$$

The solution is

$$U(x, s) = \frac{\sinh(\sqrt{s}x)}{\sinh\sqrt{s}}\left(\frac{1}{s} - \frac{1}{s + a}\right).$$

In Example 3, we already found this inverse transform:

$$\mathcal{L}^{-1}\left(\frac{\sinh(\sqrt{s}x)}{s\sinh(\sqrt{s})}\right) = x + \sum_{n=1}^{\infty}\frac{2\sin(n\pi x)}{n\pi \cos(n\pi)}e^{-n^2\pi^2 t}.$$

The inverse transform of the other term follows the same pattern. Let

$$r_a = -a: \quad q(s) = \frac{\sinh\sqrt{s}x}{\sinh\sqrt{s}}, \quad p(s) = s + a, \quad A_a(x) = \frac{\sin(\sqrt{a}x)}{\sin\sqrt{a}}$$

$$r_n = -n^2\pi^2: \quad q(s) = \frac{\sinh(\sqrt{s}x)}{s+a}, \quad p(s) = \sinh(\sqrt{s}); \quad p'(s) = \frac{1}{2\sqrt{s}}\cosh(\sqrt{s})$$

$$A_n(x) = \frac{2n\pi}{n^2\pi^2 - a}\frac{\sin(n\pi x)}{\cos(n\pi)}.$$

Therefore, the whole solution is

$$u(x,t) = x + \sum_{n=1}^{\infty} \frac{2\sin(n\pi x)}{n\pi\cos(n\pi)}e^{-n^2\pi^2 t} - \frac{\sin(\sqrt{a}x)}{\sin(\sqrt{a})}e^{-at} - \sum_{n=1}^{\infty}\frac{2n\pi}{n^2\pi^2-a}\frac{\sin(n\pi x)}{\cos(n\pi)}e^{-n^2\pi^2 t}.$$

Chapter 6

Miscellaneous Exercises

1. Transform the problem

$$U'' - \gamma^2 \left(U - \frac{T}{s} \right) = sU - T_0$$

$$U'(0) = 0, \quad U'(1) = 0$$

$$U = \frac{T_0 + \gamma^2 T/s}{\gamma^2 + s} + c_1 \cosh\left(\sqrt{\gamma^2 + s}\,x\right) + c_2 \sinh\left(\sqrt{\gamma^2 + s}\,x\right)$$

$U'(0) = 0$ makes $c_2 = 0$; $U'(1) = 0$ makes $c_1 = 0$. Thus U is just the first term. Apply partial fractions

$$\frac{T_0 + \gamma^2 T/s}{\gamma^2 + s} = \frac{T_0}{\gamma^2 + s} + \frac{\gamma^2 T}{s(\gamma^2 + s)} = \frac{T_0}{\gamma^2 + s} + T\left(\frac{1}{s} - \frac{1}{\gamma^2 + s}\right)$$

$$u(t) = (T_0 - T)e^{-\gamma^2 t} + T.$$

The partial differential equation really does not involve x, since there is nothing in the partial differential equation, the initial condition or the boundary conditions that forces u to be different for different values of x.

3. Transform the problem

$$U'' = sU, \quad U'(0) = 0, \quad U(1) = \frac{1}{s^2}.$$

Solution

$$U = \frac{\cosh(\sqrt{s}\,x)}{s^2 \cosh(\sqrt{s})}.$$

Zeros of the denominator: $s = 0$ and $s = -(n - \frac{1}{2})^2 \pi^2 = r_n$, $n = 1, 2, \cdots$. A_0: replace the cosh by the first terms of their Taylor series near 0:

$$U(x, s) \cong \frac{1 + \frac{sx^2}{2} + \cdots}{s^2 \left(1 + \frac{s}{2} + \cdots\right)} \cong \frac{\left(1 + \frac{sx^2}{2} + \cdots\right)\left(1 - \frac{s}{2} + \cdots\right)}{s^2} = \frac{1 + \frac{s}{2}(x^2 - 1) + \cdots}{s^2}$$

$$= \frac{1}{s^2} + \frac{x^2 - 1}{2s}.$$

The inverse transform of these terms is $t + \frac{x^2 - 1}{2}$ (known as a heat polynomial) which satisfies the heat equation and the boundary conditions.

A_n: Let

$$q(s) = \frac{\cosh(\sqrt{s}\,x)}{s^2}, \quad p(s) = \cosh(\sqrt{s}), \quad p' = \frac{1}{2\sqrt{s}} \sinh(\sqrt{s})$$

$$A_n = \frac{\cos((n - \frac{1}{2})\pi x)}{(n - \frac{1}{2})^4 \pi^4} \frac{2i(n - \frac{1}{2})\pi}{i\sin((n - \frac{1}{2})\pi)} = \frac{2\cos((n - \frac{1}{2})\pi x)}{(n - \frac{1}{2})^2 \pi^3 \sin((n - \frac{1}{2})\pi)}.$$

The solution is

$$u(x, t) = t + \frac{x^2 - 1}{2} + \sum_{n=1}^{\infty} \frac{2\cos(\rho_n x)}{\rho_n^3 \sin(\rho_n)} e^{-\rho_n^2 t}$$

where $\rho_n = (n - \frac{1}{2})\pi$.

5. Transform the problem:

$$U'' = sU - \frac{1}{s}, \quad U(0) = 0, \quad U(1) = 0.$$

Solution

$$U(x,s) = \frac{1}{s^2} - \frac{\sinh(\sqrt{s}(1-x)) + \sinh(\sqrt{s}x)}{s^2 \sinh(\sqrt{s})}.$$

To obtain the inverse transform of the second term easily, let $v(x,t)$ be the solution of Example 3, Section 6.3.

$$V(x,s) = \frac{\sinh(\sqrt{s}(1-x)) + \sinh(\sqrt{s}x)}{\sinh(\sqrt{s})}.$$

Therefore, using the integral rule

$$u(x,t) = t - \int_0^t v(x,t')dt' = t - \left(t + \sum_{n=1}^{\infty} \frac{2(\sin(n\pi(1-x)) + \sin(n\pi x))}{(n\pi)^3 \cos(n\pi)} \left(1 - e^{-n^2\pi^2 t} \right) \right).$$

7. Transform the problem

$$U'' = sU, \quad U(0) = 0, \quad U(1) = \frac{1}{s}$$

Solution:

$$U(x,s) = \frac{\sinh(\sqrt{s}x)}{s \sinh(\sqrt{s})}.$$

Follow the procedure of Section 6.3

$$s = 0: \quad A_0(x) = x,$$

$$s = r_n = -n^2\pi^2, \quad n = 1, 2, \cdots.$$

Choose $q(s) = \sinh(\sqrt{s}x)/s$, $p(s) = \sinh(\sqrt{s})$, $p'(s) = \cosh(\sqrt{s})/2\sqrt{s}$. Then

$$A_n(x) = \frac{q(r_n)}{p'(r_n)} = \frac{i\sin(n\pi x)/(-n^2\pi^2)}{\cos(n\pi)/(2in\pi)} = \frac{2\sin(n\pi x)}{n\pi \cos(n\pi)}$$

$$u(x,t) = x + \sum_{n=1}^{\infty} \frac{2\sin(n\pi x)}{n\pi \cos(n\pi)} e^{-n^2\pi^2 t}.$$

9. Transform the problem $U'' = sU - 1$, $U(0) = 0$, $U(x)$ bounded as $x \to \infty$.

11. Let $g(t) = 1 - \mathrm{erf}\,(x/\sqrt{4t})$. According to Exercise 10, $\mathcal{L}(g(t)) = e^{-x\sqrt{s}}/s$. Since $g(0) = 0$ [see Section 2.12], $e^{-x\sqrt{s}} = \mathcal{L}\left(\frac{dg}{dt} \right)$, and

$$\frac{dg}{dt} = -\frac{2}{\sqrt{\pi}} e^{-x^2/4t} \frac{d}{dt} \left(\frac{x}{\sqrt{4t}} \right) = \frac{e^{-x^2/4t}}{2\sqrt{\pi t^3}}.$$

13. The transform of the problem in Exercise 6 is (see the solution of 7)

$$U(x,s) = \frac{\sinh(\sqrt{s}x)}{s\sinh(\sqrt{s})} = \sum_{n=0}^{\infty} \frac{1}{s}\left(e^{-\sqrt{s}(2n+1-x)} - e^{-\sqrt{s}(2n+1+x)}\right).$$

Therefore, from Exercise 10

$$u(x,t) = \sum_{n=0}^{\infty}\left[\text{erfc}\left(\frac{2n+1-x}{\sqrt{4t}}\right) - \text{erfc}\left(\frac{2n+1+x}{\sqrt{4t}}\right)\right].$$

15. The Fourier series is $f(t) = \sum_{n=1}^{\infty}\frac{2\sin(nt)}{n}$. The term-by-term transform is

$$F(s) = \sum_{n=1}^{\infty}\frac{2}{n}\frac{n}{n^2+s^2} = \sum_{n=1}^{\infty}\frac{2}{n^2+s^2}.$$

17. The zeros of the denominator are the values of s for which $1 - e^{-2as} = 0$ or $e^{-2as} = 1$. Then $2as = 2n\pi i$ for any integer n (positive, negative or zero). Let $q(s) = G(s)$, $p(s) = 1 - e^{-2as}$, $p'(s) = 2ae^{-2as}$. Then

$$\frac{q(n\pi i/a)}{p'(n\pi i/a)} = \frac{G(n\pi i/a)}{2a}, \quad \text{and} \quad f(t) = \sum_{n=-\infty}^{\infty}\frac{1}{2a}G\left(\frac{n\pi i}{a}\right)e^{\frac{n\pi it}{a}}.$$

19. See Answers.

21.

$$G(s) = \int_0^\pi e^{-st}dt - \int_\pi^{2\pi} e^{-st}dt = \frac{1-e^{-\pi s}}{s} - \frac{e^{-\pi s} - e^{-2\pi s}}{s} = \frac{(1-e^{-\pi s})^2}{s}$$

$$F(s) = \frac{(1-e^{-\pi s})^2}{s(1-e^{-2\pi s})} = \frac{1-e^{-\pi s}}{s(1+e^{-\pi s})}.$$

23. The period is $2a = \pi$.

$$G(s) = \int_0^\pi \sin(t)e^{-st}dt = e^{-st}\frac{(-s\sin(t) - \cos(t))}{s^2+1}\Big|_0^\pi = \frac{1-e^{-\pi s}\cos(\pi)}{s^2+1}$$

$$F(s) = \frac{1+e^{-\pi s}}{(s^2+1)(1-e^{-\pi s})}.$$

25. Transform the problem:

$$U'' = s^2 U, \quad U(0) = 0, \quad U(1) = \frac{\omega}{s^2+\omega^2}.$$

$$U(x,s) = \frac{\omega}{s^2+\omega^2}\frac{\sinh(sx)}{\sinh(s)}.$$

At $s = \pm i\omega$, take $p = s^2 + \omega^2$, $p' = 2s$ and q is everything else. Then

$$s = i\omega; \quad A(x) = \frac{\omega \cdot i\sin(\omega x)}{2i\omega i\sin(\omega)} = \frac{\sin(\omega x)}{2i\sin(\omega)}$$

$$s = -i\omega; \quad A(x) = \frac{\sin(\omega x)}{-2i \sin(\omega)}.$$

The sum of these two terms transforms back into

$$\frac{\sin(\omega x)}{\sin(\omega)} \frac{e^{i\omega t} - e^{-i\omega t}}{2i} = \frac{\sin(\omega x) \sin(\omega t)}{\sin(\omega)}.$$

At $s = \pm in\pi$, take $p = \sinh(s)$, $p' = \cosh(s)$, q everything else.

At $s = in\pi$, $A_n = \dfrac{\omega}{\omega^2 - n^2\pi^2} \dfrac{i \sin(n\pi x)}{\cos(n\pi)}$

At $s = -in\pi$, $A_{-n} = \dfrac{\omega}{\omega^2 - n^2\pi^2} \dfrac{-i \sin(n\pi x)}{\cos(n\pi)}.$

The sum of these two transforms back into

$$\frac{\omega}{\omega^2 - n^2\pi^2} \frac{\sin(n\pi x)}{\cos(n\pi)} \cdot \left(ie^{in\pi t} - ie^{-in\pi t}\right) = \frac{-2\omega}{\omega^2 - n^2\pi^2} \frac{\sin(n\pi x)}{\cos(n\pi)} \cdot \sin(n\pi t).$$

27. Transform the problem:

$$U'' = U' + sU, \quad U(0) = \frac{\omega}{s^2 + \omega^2},$$

U bounded as $x \to \infty$. By the methods of Chapter 0, we find that $U(x) = c_1 e^{px} + c_2 e^{mx}$, where

$$p = \frac{1}{2} + \sqrt{s + \frac{1}{4}}, \quad m = \frac{1}{2} - \sqrt{s + \frac{1}{4}}.$$

By boundedness, we must have $c_1 = 0$, because $p > 0$ but $m < 0$. Apply the boundary condition to find c_2. See Answers.

29. Follow instructions:

$$\beta = \sqrt{\alpha^2 - \frac{1}{4}}; \quad 2\alpha\sqrt{\alpha^2 - \frac{1}{4}} = \omega.$$

In the second equation, square both sides: $4\alpha^2(\alpha^2 - \frac{1}{4}) = \omega^2$ or $4\alpha^4 - \alpha^2 - \omega^2 = 0$. Solve as a biquadratic (quadratic in α^2)

$$\alpha^2 = \frac{1 \pm \sqrt{1 + 16\omega^2}}{8}.$$

This gives the Answer.

Chapter 7

7.1 Boundary Value Problems

1. Follow the first example. Replacement equations with $n = 4$, $\Delta x = 1/4$

$$16\left(u_{i-1} - 2u_i + u_{i+1}\right) = -1, \quad i = 1, 2, 3, \quad u_0 = 0, \quad u_4 = 1.$$

These equations, written out, are

$$16\left(u_0 - 2u_1 + u_2\right) = -1 \quad (i = 1)$$

$$16\left(u_1 - 2u_2 + u_3\right) = -1 \quad (i = 2)$$

$$16\left(u_2 - 2u_3 + u_4\right) = -1 \quad (i = 3)$$

Now replace u_0 and u_4 by the specified values and simplify:

$$-32u_1 + 16u_2 \qquad = -1$$

$$16u_1 - 32u_2 + 16u_3 = -1$$

$$16u_2 - 32u_3 = -17$$

This system of 3 equations in 3 unknowns can be solved by elimination to find

$$u_1 = \frac{11}{32}, \qquad u_2 = \frac{5}{8}, \qquad u_3 = \frac{27}{32}.$$

3. Follow the first example. Replacement equations with $n = 4$, $\Delta x = 1/4$:

$$16\left(u_{i-1} - 2u_i + u_{i+1}\right) - u_i = -\frac{i}{2}, \quad i = 1, 2, 3, \quad u_0 = 0, \quad u_4 = 1.$$

Write out the equations:

$$16\left(u_0 - 2u_1 + u_2\right) - u_1 = -\frac{1}{2} \quad (i = 1)$$

$$16\left(u_1 - 2u_2 + u_3\right) - u_2 = -1 \quad (i = 2)$$

$$16\left(u_2 - 2u_3 + u_4\right) - u_3 = -\frac{3}{2} \quad (i = 3)$$

Replace u_0 and u_4 by their given values and simplify:

$$-33u_1 + 16u_2 \qquad = -\tfrac{1}{2}$$

$$16u_1 - 33u_2 + 16u_3 = -1$$

$$16u_2 - 33u_3 = -\tfrac{35}{2}$$

Solve by elimination (use software)

$$u_1 = .2849, \quad u_2 = .5563, \quad u_3 = .8000.$$

5. Replacement equations with $n = 4$:

$$16 \left(u_{i-1} - 2u_i + u_{i+1} \right) = \frac{i}{4}, \quad i = 0, 1, 2, 3.$$

Note that $i = 0$ is included because the boundary condition at $x = 0$ does not specify $u(0)$: it contains a derivative. The replacement for the condition at $x = 0$ is

$$u_0 - \frac{u_1 - u - 1}{2 \cdot 1/4} = 1$$

from which $u_{-1} = u_1 + 1/2 - u_0/2$. The replacement equation with $i = 0$ is

$$u_{-1} - 2u_0 + u_1 = 0.$$

Substitute for u_{-1}:

$$u_1 + \frac{1}{2} - \frac{u_0}{2} - 2u_0 + u_1 = 0 \ \text{ or } \ -\left(\frac{5}{2} \right) u_0 + 2u_1 = -\frac{1}{2} \ \ (i = 0).$$

The replacement equations for $i = 1, 2, 3$, with $u_4 = 0$, are

$$
\begin{aligned}
16u_0 - 32u_1 + 16u_2 \quad\quad &= \tfrac{1}{4} \ \ (i = 1) \\
16u_1 - 32u_2 + 16u_3 &= \tfrac{2}{4} \ \ (i = 2) \\
16u_2 - 32u_3 &= -\tfrac{3}{4} \ \ I = 3)
\end{aligned}
$$

Now the four equaitons are solved simultaneously (with software) to find

$$u_0 = .4219, \quad u_1 = .2773, \quad u_2 = .1484, \quad u_3 = .0508.$$

7. Replacement equations in either case are

$$\frac{1}{(\Delta x)^2} \left(u_{i-1} - 2u_i + u_{i+1} \right) + 10u_i = 0, \quad i = 1, 2, \cdots, n - 1,$$

$$u_0 = 0, \quad u_n = -1.$$

With $n = 3, \Delta x = 1/3, 1/(\Delta x)^2 = 9$:

$$-8u_1 + 9u_2 = 0$$

$$9u_1 - 8u_2 = 0$$

Solution: $u_1 = 4.765, u_2 = 4.235$.

With $n = 4, \Delta x = 1/4, 1/(\Delta x)^2 = 16$:

$$
\begin{aligned}
-22u_1 + 16u_2 \quad\quad &= 0 \\
16u_1 - 22u_2 + 16u_3 &= 0 \\
16u_2 - 22u_3 &= 16
\end{aligned}
$$

Solution: $u_1 = 6.649, u_2 = 9.143, u_3 = 5.922$.

The values are not directly comparable, but a sketch shows that the results differ violently. See Answers.

9. Replacement equation for $n = 5$:

$$25 \left(u_{i-1} - 2u_i + u_{i+1} \right) - 25u_i = -25, \quad i = 1, 2, 3, 4, 5. \tag{$*$}$$

$$u_0 = 2; \quad u_5 + 5(u_6 - u_4) = 1.$$

Note that the replacement equation for the differential equation has to be included at $i = 5$, because the second boundary condition includes a derivative. There are two ways to proceed:

Procedure 1. Solve the replacement for the second boundary condition for u_6:

$$u_6 = u_4 - \frac{u_5}{5} + \frac{1}{5}.$$

Substitute into (*) for $i = 5$:

$$25(u_4 - 2u_5 + u_6) - 25u_5 = -25$$

to get $50u_4 - 80u_5 = -30$. The system to solve is

$$-75u_1 + 25u_2 \qquad\qquad\qquad\qquad = -25$$
$$25u_1 - 75u_2 + 25u_3 \qquad\qquad\qquad = -25$$
$$25u_2 - 75u_3 + 25u_4 \qquad\qquad = -25$$
$$25u_3 - 75u_4 + 25u_5 = -25$$
$$50u_4 - 80u_5 = -30$$

Procedure 2. Use (*) for $i = 1$ to 5 replacing u_0 by 2. Add on the *replacement* for the second boundary condition. The system to solve is

$$-75u_1 + 25u_2 \qquad\qquad\qquad\qquad\qquad = -25$$
$$25u_1 - 75u_2 + 25u_3 \qquad\qquad\qquad\qquad = -25$$
$$25u_2 - 75u_3 + 25u_4 \qquad\qquad\qquad = -25$$
$$25u_3 - 75u_4 + 25u_5 \qquad\qquad = -25$$
$$25u_4 - 75u_5 + 26u_6 = -25$$
$$-5u_4 + u_5 + 5u_6 = -1$$

The solution of this system produces a value for u_6, which is discarded. Otherwise both procedures give the same results. As long as software does the work, the second procedure is smoother.

11. Replacement equations:

$$\frac{1}{(\Delta x)^2} (u_{i-1} - 2u_i + u_{i+1}) + \frac{1}{2\Delta x} (u_{i+1} - u_{i-1}) - u_i = -i\Delta x, \quad i = 0, 1, 2, \cdots, n$$

$$\frac{1}{\Delta x}(u_1 - u_{-1}) = 0, \quad u_n = 1.$$

Now set $n = 3$, $\Delta x = 1/3$

$$9(u_{-1} - 2u_0 + u_1) + \left(\frac{3}{2}\right)(u_1 - u_{-1}) - u_0 = 0$$

$$9(u_0 - 2u_1 + u_2) + \left(\frac{3}{2}\right)(u_2 - u_0) - u_1 = 0$$

$$9(u_1 - 2u_2 + u_3) + \left(\frac{3}{2}\right)(u_3 - u_1) - u_2 = 0$$

Note that $u_{-1} = u_1$ and $u_3 = 1$, so these simplify to

$$
\begin{aligned}
-19u_0 + \quad 18u_1 \quad\quad\quad\quad &= \quad 0 \\
(15/2)u_0 - \quad 19u_1 + (21/2)u_2 &= \ -1/3 \\
(15/2)u_1 - \quad 19u_2 &= -67/6
\end{aligned}
$$

Solution $u_0 = .795$, $u_1 = .839$, $u_2 = .919$.

Chapter 7

7.2 Heat Problems

1. The replacement equations are as in Equation (8) and (since $r = 1/2$) Equation (9). The line for $m = 0$ would be

$$0 \qquad .25 \qquad .5 \qquad .75 \qquad 0$$

which is exactly the line for $m = 1$ in Table 4.

3. The rule of thumb is that all coefficients in Equation (13) must be ≥ 0. Thus $1 - 2r \geq 0$ and $1 - 2r - \frac{1}{2}r\gamma = 1 - 2.5r \geq 0$. The stricter requirement is the second, so $r \leq 0.4$. See Answers.

5. The replacement equations are as in Equation (8), so the maximum value of r is $1/2$, $\Delta t = 1/32$. Then you compute using Equation (9), but with $u_0(m) = u_4(m) = m\Delta t = m/32$. See Answers.

7. The replacement equations are

$$\frac{u_{i-1}(m) - 2u_i(m) + u_{i+1}(m)}{(\Delta x)^2} = \frac{u_i(m+1) - u_i(m)}{\Delta t} - 1.$$

From here, we obtain

$$u_i(m+1) = ru_{i-1}(m) + (1 - 2r)u_i(m) + ru_{i+1}(m) + \Delta t.$$

Evidently the maximum value for r is $1/2$, so $\Delta t = 1/32$. Then the equation to be used for computing is

$$u_i(m+1) = \frac{1}{2}\left(u_{i-1}(m) + u_{i+1}(m)\right) + \frac{1}{32}.$$

See Answers.

9. The boundary condition at $x = 0$ is replaced by

$$\frac{u_{+1}(m) - u_{-1}(m)}{2\Delta t} = 0$$

or $u_{-1}(m) = u_1(m)$. Then, following Equations (7) and (8), the replacement at $x = 0$ ($i = 0$) is

$$u_0(m+1) = ru_{-1}(m) + (1 - 2r)u_0(m) + ru_1(m) = (1 - 2r)u_0(m) + 2ru_1(m)$$

using the boundary condition. The rest of the replacement equations are as in Equation (8). Hence, $r = 1/2$ is the maximum value of r. Using this the equations for calculation are $u_0(m+1) = u_1(m)$ and the three equations of Equation (9). See Answers.

Chapter 7

7.3 Wave Equation

1. The replacement equations are Equation (4) with $\rho = 1$:

$$u_i(m+1) = u_{i-1}(m) + u_{i+1}(m) - u_i(m-1)$$

for $i = 1, 2, 3$ and $m = 1, 2, \cdots$. The boundary conditions give $u_0(m) = 0$, $u_4(m) = 0$. The start-up equation in Equation (7), with $f(x) \equiv 0$, $g(x) \equiv 1$ and $\Delta t = 1/4$, it becomes $u_i(1) = 1/4$, $i = 1, 2, 3$. See Answers.

3. The setup is the same as in Exercise 1 above, except that the start-up equation is

$$u_i(1) = \frac{1}{4} \sin(i\pi\Delta x).$$

That is, $u_1(1) = \frac{1}{4}\sin(\frac{\pi}{4}) = \sqrt{2}/8 = .177$, $u_2(1) = \frac{1}{4}\sin(\pi/2) = \frac{1}{4}$, $u_3(1) = u_1(1)$.

5. The replacement equations are in Equation (4), with $\rho^2 = 1/2$:

$$u_i(m+1) = \frac{1}{2}u_{i-1}(m) + u_i(m) + \frac{1}{2}u_{i+1}(m) - u_i(m-1)$$

for $i = 1, 2, 3$ and $m = 1, 2, 3, \cdots$. The starting equation is Equation (7), with $g(x) \equiv 0$:

$$u_i(1) = \frac{1}{4}f(x_{i-1}) + \frac{1}{2}f(x_i) + \frac{1}{4}f(x_{i+1}).$$

Specifically, $u_1(1) = u_3(1) = \frac{1}{2}$, $u_2(1) = \frac{3}{4}$. See Answers.

The periodicity in time will not show as clearly here as when $\rho = 1$, because the period in time, 2, is not an integer multiple of $\Delta t = \rho\Delta x = 1/4\sqrt{2} = .177$.

7. The replacement equations are as in the solution of Exercise 1, above, with the note that $u_0(m) = 0$ and $u_4(m) = h(m/4)$. From the specifications of h, we have $h(m/4) = 1$ for $m = 1, 2, 3$, $h(1) = 0$, $h(m/4) = -1$ for $m = 5, 6, 7$, $h(2) = 0$, and h has period 2. Because the function $h(t)$, which moves the right-hand boundary value, has the same period as a product solution, the values of u will build up over time.

9. First replace the partial derivatives as in the beginning of the section

$$\frac{u_{i+1}(m) + 2u_i(m) + u_{i-1}(m)}{(\Delta x)^2} = \frac{u_i(m+1) - 2u_i(m) + u_i(m-1)}{(\Delta t)^2} + 16u_i(m).$$

Solve for $u_i(m+1)$ and simplify. Here is the first step: multiply through by Δt^2

$$u_i(m+1) - 2u_i(m) + u_i(m-1) + 16\Delta t^2 u_i(m) = \rho^2\left(u_{i+1}(m) - 2u_i(m) + u_{i-1}(m)\right).$$

The final simplification is

$$u_i(m+1) = (2 - 2\rho^2 - 16\Delta t^2)u_i(m) + \rho^2 u_{i+1}(m) + \rho^2 u_{i-1}(m) - u_i(m-1).$$

Use the relation $\rho^2 = (\Delta t/\Delta x)^2 = 16(\Delta t)^2$ to rewrite the coefficient of $u_i(m)$ as $2 - 3\rho^2$. This must not be negative, so the largest acceptable value of ρ makes $\rho^2 = 2/3$ or $\rho = \sqrt{2/3}$, and $\Delta t = \rho\Delta x = \sqrt{2/3}/4 = 1/\sqrt{24} = .204$. Probably choosing $\Delta t = .2$ would make arithmetic easier.

Chapter 7

7.4 Potential Equation

1. The replacement equations come from Equation (2)

$$\frac{u_N + u_S + u_E + u_W - 4u_i}{\Delta x^2} = -1$$

or, using $\Delta x = 1/4$,

$$u_i = \frac{u_N + u_S + u_E + u_W + \frac{1}{16}}{4}.$$

Look at Figure 2 for numbering points, and recall that $u = 0$ at all boundary points. Symmetry gives: $u_1 = u_3 = u_7 = u_9$; $u_2 = u_4 = u_6 = u_8$. Thus, there are only three different values to find, with these equations:

$$u_1 = \frac{2u_2 + \frac{1}{16}}{4}$$

$$u_2 = \frac{2u_1 + u_5 + \frac{1}{16}}{4}$$

$$u_5 = \frac{4u_2 + \frac{1}{16}}{4}.$$

The matrix form of the system is

$$\begin{bmatrix} 1 & -\frac{1}{2} & 0 \\ -\frac{1}{2} & 1 & -\frac{1}{4} \\ 0 & -1 & 1 \end{bmatrix} \begin{bmatrix} u_1 \\ u_2 \\ u_5 \end{bmatrix} = \frac{1}{64} \begin{bmatrix} 1 \\ 1 \\ 1 \end{bmatrix}$$

3. The replacements are routine. Use the numbering of Figure 2 and note that $u_2 = u_4$, $u_3 = u_7$, $u_6 = u_8$, leaving six unknowns. The equations are

$$-4u_1 + 2u_2 = 0$$
$$u_1 - 4u_2 + u_3 + u_5 = 0$$
$$u_2 - 4u_3 + u_6 = -1/4$$
$$2u_2 - 4u_5 + 2u_6 = 0$$
$$u_3 + u_5 - 4u_6 + u_9 = -1/2$$
$$2u_6 - 4u_9 = -3/2$$

5. See Figure 5. There is symmetry left to right (e.g., $u_7 = u_{12}$), top to bottom (e.g., $u_7 = u_{21}$) and across the line $y = x$ (e.g., $u_7 = u_2$). As a result, only five unknowns must be found. Replacements follow Equation (7).

$$-4u_1 + 2u_2 = 0$$
$$u_1 - 4u_2 + u_3 + u_8 = 0$$
$$u_2 - 3u_3 + u_9 = 0$$
$$2u_2 - 4u_8 + 2u_9 = 0$$
$$u_3 + u_8 - 3u_9 = -1$$

In the third equation, the coefficient of u_3 is -3 because $u_4 = u_3$. Similarly in the fifth equation.

7. The replacements come directly from Equation (2). Note that $1/(\Delta x)^2 = 25$. The typical equation (similar to Equation (7)) is

$$u_N + u_S + u_E + u_W - 4u_i = -1.$$

By symmetry, $u_3 = u_7$, $u_4 = u_8$, $u_5 = u_9$, $u_6 = u_{10}$; only u_i for $i = 1..6$ need to be found. The equations are

$$-4u_1 + u_2 + u_3 = -1$$
$$u_1 - 4u_2 + u_4 = -1$$
$$u_1 - 3u_3 + u_4 = -1$$
$$u_2 + u_3 - 3u_4 + u_5 = -1$$
$$u_4 - 3u_5 + u_6 = -1$$
$$u_5 - 3u_6 = -1$$

9. There is left-right symmetry, so only u_i for $i = 1, 2, 3, 4$ need to be found. Note that each point has at most two neighbors inside the region, and $u_4 = u_5$. The equations are:

$$-4u_1 + u_2 = -1$$
$$u_1 - 4u_2 + u_3 = -1$$
$$u_2 - 4u_3 + u_4 = -2$$
$$u_3 - 3u_4 = -1$$

Chapter 7

7.5 Two-Dimensional Problems

1. Generally, the replacement equations are the same as Equation (8), so the largest stable value of r is 1/4 ($\Delta t = r\Delta x^2 = 1/64$). Using this value and the given boundary conditions we find these replacement equations

$$u_1(m+1) = \frac{u_2(m) + u_4(m)}{4}$$

$$u_2(m+1) = \frac{u_1(m) + u_3(m) + u_5(m)}{4}$$

$$u_3(m+1) = \frac{u_2(m) + u_6(m)}{4}$$

$$u_4(m+1) = \frac{u_1(m) + u_5(m) + 1}{4}$$

$$u_5(m+1) = \frac{u_2(m) + u_4(m) + u_6(m) + 1}{4}$$

$$u_6(m+1) = \frac{u_3(m) + u_5(m) + 1}{4}$$

Initially all values are 0.

3. The replacement equations follow from Equation (8), and the largest stable r is 1/4. Because the geometry, the boundary conditions and the initial condition are all symmetric, we must have $u_1 = u_2 = u_4 = u_5$. Thus the equations simplify to

$$u_1(m+1) = \frac{u_3(m)}{4}$$

$$u_3(m+1) = u_1(m)$$

5. See Answers.

7. See Answers. The starting equation comes from Equation (22) and (23). With $f \equiv 0$, this is $u_i(1) = \Delta t g_i$; that is, using the numbering from Figure 2, $u_5(1) = 4\sqrt{2}\Delta t = 1$ and $u_i(1) = 0$ at other points.

9. See Answers. The starting equations reduce to

$$u_1(1) = u_3(1) = u_7(1) = u_9(1) = 1/2$$

$$u_2(1) = u_4(1) = u_6(1) = u_8(1) = 3/4$$

$$u_5(1) = 1$$

11. See Answers for point numbering. The starting equations reduce to $u_2(1) = u_3(1) = 1/4$, and $u_i(1) = 0$ for other indices. The running equations are from Equation (21):

$$u_1(m+1) = \frac{1}{2}(u_2(m) + u_3(m)) - u_1(m-1)$$

$$u_2(m+1) = \frac{1}{2}(u_1(m) + u_4(m)) - u_2(m-1)$$

$$u_3(m+1) = \frac{1}{2}(u_1(m) + u_4(m) + u_5(m)) - u_3(m-1)$$

$$u_4(m+1) = \frac{1}{2}(u_2(m) + u_3(m) + u_6(m)) - u_4(m-1)$$

The remaining points have equations that follow the pattern of those for u_3 and u_4

$$u_{2k-1}(m+1) = \frac{1}{2}(u_{2k-3}(m) + u_{2k}(m) + u_{2k+1}(m)) - u_{2k-1}(m-1)$$

$$u_{2k}(m+1) = \frac{1}{2}(u_{2k-2}(m) + u_{2k-1}(m) + u_{2k+2}(m)) - u_{2k}(m-1)$$

Chapter 7

Miscellaneous

1. See Answers.

3. See Answers.

5. Follow procedures from Section 7-2. With $\Delta x = 1/4$ and $\Delta t = 1/32$, we have $r = 1/2$. The equations to use are as in Equation (9) of 7-2. The boundary conditions are $u_0(m) = u_4(m) = 1 - e^{-m/32}$.

7. See Answers for replacement equations. Note that $r = \Delta t/(\Delta x)^2 = 1/3$. The solution of the second problem goes to 0 much faster than the first.

9. If $\Delta x = 1/5$ and $r = 1/2$, then $\Delta t = 1/50$. The replacement equations are as in Section 7.2, Equation (9). The left boundary condition is $u_0(m) = 25m\Delta t = m/2$. See Answers.

11. The replacement equations are Equation (4) from Section 7.3 with $\rho = 1$. The boundary condition on the right is $u_4(m) = 1$ for $m \geq 1$. The starting equation, Equation (7) in Section 7.3, gives $u_i(1) = 0$ for $i = 1, 2, 3$.

13. The replacement equations come from Equation (7) of Section 7.4. For numbering, let u_{ij} be the approximation to $u(i/4, j/4)$. The boundary conditions are

$$u_{41} = .156, \quad u_{42} = .295, \quad u_{43} = .410, \quad u_{44} = .5;$$

$$u_{14} = .844, \quad u_{24} = .705, \quad u_{34} = .590;$$

$$u_{i0} = 0, \quad i = 1, 2, 3; \quad u_{0j} = 1, \quad j = 1, 2, 3.$$

15. Replacement equations come from Equation (8) of Section 7.5. Since $r = 1/4$, $\Delta t = 1/64$. The initial condition is $u = 1$ everywhere on the interior of the square. Use Figure 2 of Section 7.4 for numbering. By symmetry, only $u_1(m)$, $u_2(m)$ and $u_5(m)$ need to be found.

Printed and bound by CPI Group (UK) Ltd, Croydon, CR0 4YY

03/10/2024

01040310-0015